WITHDRAWN

Ecology of Marine Parasites

AUSTRALIAN ECOLOGY SERIES

General Editor: Harold Heatwole

Other titles in series

Reptile Ecology: Harold Heatwole
Insect Ecology: E. G. Matthews

Ecology of Marine Parasites

Klaus Rohde

University of Queensland Press
St Lucia · London · New York

© University of Queensland Press, St Lucia, Queensland 1982

This book is copyright. Apart from any fair dealing for the purposes of private study, research, criticism, or review, as permitted under the Copyright Act, no part may be reproduced by any process without written permission. Enquiries should be made to the publishers.

**Typeset by Watson Ferguson & Co., Brisbane
Printed and bound by Silex Enterprise & Printing Co., Hong Kong**

Distributed in the United Kingdom, Europe, the Middle East, Africa, and the Caribbean by Prentice-Hall International, International Book Distributors Ltd, 66 Wood Lane End, Hemel Hempstead, Herts., England.

*National Library of Australia
Cataloguing-in-Publication data*

Rohde, Klaus, 1932-.
   Ecology of marine parasites.

   Includes index.
   ISBN 0 7022 1660 7.
   ISBN 0 7022 1670 4 (pbk.).

1. Parasitism. 2. Marine biology. 3. Marine ecology.
I. Title. (Series: Ecology series ISSN 0156-6695).

574.5'249

Library of Congress Cataloguing in Publication Data

Rohde, Klaus, 1932-.
   Ecology of marine parasites

   (Australian ecology series)
   Includes bibliographical references and indexes.
   1. Parasites—Ecology. 2. Marine ecology. I. Title.
II. Title: Marine parasites. III. Series.
QL757.R63   574.5'2636   81-12934
ISBN 0-7022-1660-7   AACR2
ISBN 0-7022-1670-4 (pbk.)

Contents

List of Figures   *ix*
Black and White   *ix*
Colour   *xi*
List of Tables   *xiii*
Preface: Ecology in Australia   *xv*
Introduction   *1*

CHAPTER 1   **The Nature of Parasitism**   *4*
    Definition of parasitism; types of parasites and hosts; other associations   *4*
    Ecological, medical and economic importance of parasites in general   *5*
    Present state of marine parasitology   *6*

CHAPTER 2   **The Kinds of Marine Parasites**   *8*

CHAPTER 3   **The Kinds of Hosts of Marine Parasites**   *47*

CHAPTER 4   **Parasites of Parasites**   *50*

CHAPTER 5   **General Adaptations of Parasitic Animals**   *52*
    Size of parasites   *52*
    Increase in reproductive capacity   *54*
    Two opposing trends: reduction and increase in complexity   *55*
    Dispersal   *59*
    Mechanisms of infection   *61*
    Aggregation   *65*
    Hermaphroditism, parthenogenesis and asexual reproduction   *67*
    "Goodness" of parasitism   *69*

CHAPTER 6 Host–Parasite Interactions 72
Behavioural reactions of hosts to parasites 72
Immunity and tissue reactions 86
Effects on host individuals 90
Effects on host populations; mass mortalities 96
Co-evolution of hosts and parasites; use of parasites for the study of host phylogeny and origin 100

CHAPTER 7 The Ecological Niches of Parasites 104
The niche concept 104
Niche dimensions of parasites 104
Saturation of niches with species 116
Causes of niche restriction 118
Biological functions of niche restrictions 127

CHAPTER 8 Characteristics of Parasite Faunas of Different Seas 136
Zoogeographical regions 136
Differences in the parasite faunas of the Atlantic and Indo-Pacific Oceans 137
Latitudinal gradients in species diversity of hosts and parasites 140
Latitudinal gradients in frequencies and intensities of infection 141
Latitudinal gradients in niche width 144
Fluctuations of parasite infections in cold and warm seas 147
Differences between parasite faunas in shallow and deep water 148
Relict parasite faunas in isolated seas and hosts 150
Parasite endemicity of remote oceanic islands 151
Importance of temperature for parasite distribution 151
Importance of age of oceans for parasite diversity 154
Effects of host migrations on parasites; parasites as biological markers 155

CHAPTER 9 Economic and Hygienic Importance of Marine Parasites 157
Parasitic diseases of marine fishes 157
Parasitic diseases of marine molluscs 160
Parasitic diseases of marine crustaceans 163

Parasitic diseases of marine mammals 164
Economic importance of marine parasites 166
Control of parasitic diseases of marine animals 168
Human infections with marine parasites 170

CHAPTER 10 Future Research *174*
Taxonomy *174*
Species formation *174*
Phylogeny *175*
Zoogeography *175*
Ecology *176*
Marine parasites as biological control agents *176*

APPENDIX The Animal Kingdom *178*
GLOSSARY *181*
REFERENCES *193*
INDEXES *231*

Figures

BLACK AND WHITE

1. Phylogeny of animal phyla   9
2. *Syndinium turbo* (Sarcomastigophora)   *13*
3. *Ellobiopsis chattoni* (Sarcomastigophora)   *13*
4. *Trypanosoma rajae* (Sacromastigophora)   *13*
5. *Haemogregarina delagei* (Apicomplexa)   *13*
6. *Aggregata octopiana* (Apicomplexa)   *13*
7. Oocyst of *Eimeria clupearum* (Apicomplexa)   *13*
8. Life cycle of *Aggregata eberthi* (Apicomplexa)   *15*
9. *Hypocomella cardii* (Ciliophora)   *17*
10. *Caliperia brevipes* (Ciliophora)   *17*
11. *Trichodina parabranchicola* (Ciliophora)   *17*
12. *Oenophorachona ectenolaemus* (Ciliophora)   *17*
13. *Protoopalina saturnalis* (Sarcomastigophora)   *17*
14. Development of *Tachyblaston ephelotensis* (Ciliophora)   *18*
15. *Myxidium coryphaenoidium* (Myxozoa)   *19*
16. *Minchinia nelsoni* (Ascetospora)   *19*
17. *Rhopalura* sp. (Mesozoa)   *19*
18. *Dicyema hypercephalum* (Mesozoa)   *19*
19. *Polypodium hydriforme* (Cnidaria)   *22*
20. *Gastrodes* sp. (Ctenophora)   *22*
21. *Ectocotyla paguri* (Turbellaria)   *22*
22. *Fecampia xanthocephala* (Turbellaria)   *22*
23. Life cycle of digenetic trematode   *24*
24. Life cycle of *Lobatostoma manteri* (Aspidogastrea)   *25*
25. *Sprostonia longiphallus* (Monogenea, Monopisthocotylea)   *27*
26. *Heteromicrocotyla australiensis* (Monogenea, Polyopisthocotylea)   *27*
27. *Gyrocotyle* sp. (Cestodaria)   *28*

x List of Figures

28. Larval trypanorhynch (Eucestoda) 28
29. *Gononemertes* (Nemertina) 30
30. Life cycle of *Thynnascaris aduncum* (Nematoda) 30
31. *Zelinkiella synaptae* (Rotatoria) 30
32. *Nectonema*, adult (Nematomorpha) 32
33. *Nectonema*, larva (Nematomorpha) 32
34. Larval Acanthocephala 32
35. *Crangonobdella murmanica* (Hirudinea) 34
36. *Myzostoma* sp. (Myzostomida) 34
37. *Ichthyotomus sanguinarius* (Polychaeta) 34
38. Larval Pycnogonida 36
39. *Phoxichilidium femoratum* (Pycnogonida) 36
40. *Caligus rapax* (Copepoda) 36
41. *Argulus* (Branchiura) 38
42. *Dendrogaster murmanensis* (Ascothoracida) 38
43. *Sacculina carcini* (Rhizocephala) 40
44. *Praniza* larva of *Gnathia* sp. (Isopoda) 40
45. *Cancricepon elegans* (Isopoda) 41
46. *Portunion maenadis* (Isopoda) 41
47. *Danalia curvata* (Isopoda) 41
48. *Cyamus erraticus* (Amphipoda) 41
49. *Tetrakentron synaptae* (Tardigrada) 43
50. *Rhopalomenia aglaopheniae* (Solenogastres) 43
51. *Thyca stellasteris* (Gastropoda) 43
52. *Entoconcha mirabilis* (Gastropoda) 43
53. *Carapus acus* (Teleostei) 46
54. *Edriolychnus schmidti* (Teleostei) 46
55. *Crassicauda crassicauda* (Nematoda) in fin whale 53
56. Nervous system of *Diaschistorchis multitesticularis* (Digenea) 57
57. Opisthaptor of *Entobdella soleae* (Monogenea) 58
58. Marine cercariae 60
59. Cleaning symbioses 78
60. Host specificity of marine Monogenea and trematodes at different latitudes 109
61. Microhabitats of parasites in digestive tract of flounder 110
62. Microhabitats of ectoparasites of mackerel 111
63. Attachment of cestode to intestinal mucosa of fish 121
64. Distribution of cysts and small Monogenea on gills of four species of fish 134

65. Relative species diversity of Monogenea and trematodes at different latitudes in the Pacific and Atlantic Oceans *135*
66. Latitudinal gradients in species diversity of marine fishes *138*
67. Frequencies of infection with Monogenea at different latitudes *143*
68. Host ranges of marine Monogenea and trematodes at different latitudes *144*
69. Microhabitats of ectoparasites on marine fishes *145*
70. Effect of parasite numbers on microhabitat width *146*
71. Effect of nematode infection on cod liver *159*

COLOUR

Plate I–VI following *page 71*

Plate I Some parasitic protozoans, mesozoans and cestodes
1. Spores of the myxozoan *Kudoa*
2. The mesozoan *Dicyema*
3. A trypanorhynch cestode, tentacles retracted
4. A trypanorhynch cestode, tentacles protracted
5. Tentacle of a trypanorhynch cestode
6. Cestode head from carpet shark

Plate II Some marine trematodes
1. *Staphylorchis* from the body cavity of a shark
2. *Opecoelina* from the intestine of bream
3. *Nasitrema* from nasal cavity of dolphin
4. The schistosome *Austrobilharzia* from the blood vessels of silver gulls
5. The didymozoid *Nemathobothrium* from mackerel
6. An immature didymozoid

Plate III Some marine trematodes
1. *Syncoelium* from the gill rakers of skipjack tuna
2. Ventral sucker of *Syncoelium* attached to spines of gill rakers
3. The aspidogastrid *Rugogaster* from chimaeras
4. The aspidogastrid *Lobatostoma* in the digestive gland of snail (section)
5. The aspidogastrid *Lobatostoma* in the stomach of snail (section)
6. Uninfected stomach of snail (section)

Plate IV   Some marine monogeneans
1. *Eurysorchis* from gill arches of warehou
2. Clamp of *Eurysorchis*
3. *Kahawaia* from gill filaments of Australian salmon
4. Posterior attachment organ of *Kahawaia*
5. A capsalid from mackerel tuna
6. An ancyrocephaline from red mullet

Plate V   Some marine parasitic crustaceans
1. Praniza larva (Isopoda) from gills of samson fish
2. Isopod from mouth cavity of scad
3. The copepod *Pseudocycnoides* from gills of broad-barred king mackerel
4. Head of *Pseudocycnoides*
5. The copepod *Caligus*
6. Ventral view of *Caligus*

Plate VI   A gyrocotylid, and marine hyperparasites and cleaners
1. *Gyrocotyle* from chimaera
2. Hyperparasitic monogeneans, *Udonella*, on copepod *Caligus*
3. *Udonella* on *Caligus*, more strongly enlarged
4. Cleaner wrasse, *Labroides*, cleaning coral trout.
5. A large cod, *Epinephelus tauvina*, with two cleaner fish, *Labroides dimidiatus* (photo. W. Deas)
6. Grabham's cleaner shrimp, *Hippolysmata grabhami* (photo. W. Deas)

Cover: A large cod, *Epinephelus tauvina*, with two cleaner fish, *Labroides dimidiatus* (photo. W. Deas)

ERRATA

The first paragraph of Chapter 2 has been set incorrectly. It should read:

> Parasites are smaller than their hosts and more difficult to study. We must assume, therefore, that the relative number of parasitic species is even larger than it seems at present. But even the data presently available show how widespread and common the parasitic way of life is. Seven phyla, some of them very large, consist entirely of parasites, and many other phyla contain a large proportion of parasites. Many of these parasites are marine, as indicated for instance by the lists of marine parasites in Florida compiled by Hutton and Sogandares-Bernal (1960) and Hutton (1964).

Two illustrations in the colour section have been printed reverse image, they are the second illustration in Plate IV and the fourth illustration in Plate V.

Cover Illustration: the head of a trypanorhynch cestode.

Tables

1. Number of parasite species among various groups of animals *10*
2. Fecundity in free-living and parasitic flatworms *54*
3. Number of cleaners in various animal groups *74*
4. Host specificity of parasites in the Barents Sea *106*
5. Genera of Monogenea and trematodes in the Atlantic and Indo-Pacific *139*
6. Latitudinal gradients in monogenean and trematode genera *141*
7. Numbers of species of trematodes in different seas *142*
8. Helminths of seals and whales at different latitudes *143*
9. Endemicity of marine fishes and their parasites at oceanic islands *152*

Preface: Ecology in Australia

Ecology is the science involved with the interactions of organisms and their physical and biotic environments. This field has always been a source of fascination to professional biologists, naturalists and conservationists. In recent years, as human population has progressively increased, environmental problems have also become of vital interest and importance to the public as well. It has now become imperative that ecological principles, and the ecology of specific regions, be understood by a wide variety of people. The present series is designed to help fill this need.

It is felt that the volumes in this series will serve as a source of information for university students, teachers and the interested public who require a basic factual knowledge to broaden their understanding of ecology, and for those conservationists, agriculturists, foresters, wild life officers, politicians, engineers, etc. who may need to apply ecological principles in solving specific environmental problems. In addition, it is hoped that the series will be a valuable reference work and source of stimulation for professional ecologists, botanists and zoologists.

The study of ecology can be approached on various levels. For example, one can emphasize the biotic community and analyse the kinds and numbers of organisms living together in a particular habitat, the way they are organized in space and time and the interactions they have with each other. This type of ecology is known as synecology.

Another way of studying ecology is by systems analysis. In this method the biotic community and the physical environment, which together make up what is known as an ecosystem, are looked upon as a functioning unit. In such an approach the main emphasis is on the cycling of energy, minerals or organic materials within the ecosystem and the factors influencing these processes, rather than specifically upon the organisms themselves. Often mathematical or theoretical models are constructed and tested, frequently with the aid of computers.

Both of the above approaches are synthetic; they take an overview of entire communities or systems and do not emphasize individual species. By contrast the following two approaches, collectively known as autecology, are concerned mainly with particular species.

The population approach, often called demography, is concerned with: (1) fluctuation in the abundance and distribution of individuals of a given species in an area; (2) the contributing phenomena such as birth and death rates, immigration, emigration, longevity and survival; and (3) the influence of the physical environment and of other species on these characteristics. Of major interest are mechanisms regulating population density and factors influencing population stability.

The final approach to ecology is one primarily concerned with the effect of the environment on the individuals of a species, that is, how they are affected by temperature, moisture, light or other external factors. This approach is known variously as environmental physiology or physiological ecology. The keynote is adaptation to specific environments.

All of the above approaches are employed with varying emphasis in the volumes of this series.

Certain topics, such as ecology of grasslands, ecology of forests and woodlands, or ecology of deserts lend themselves to a community approach; grassland, forest and desert *are* types of communities and if studied as an entity must be approached on the community or ecosystem level. On the other hand, where specific taxa such as reptiles, birds or mammals are treated, the autecological approach is more often used. The particular aspect emphasized varies from group to group, depending on the information available.

Regardless of emphasis, in each book of this series the available information in a particular field is reviewed critically and summarized, so that the reader might be brought abreast of current knowledge and developments. Recent trends are indicated and the foundations for future developments are prepared by highlighting conspicuous gaps in knowledge and pointing out what appear to be fruitful avenues for research.

Harold Heatwole

Armidale, N.S.W.
March 1981

Introduction

Ecology is the study of the relationship between living organisms and their environments. The total environment of an animal or plant is complex and consists not only of its physical world, such as the soil, rocks and water of its habitat and the temperature conditions, humidity, wind and oxygen levels that surround it, but its biotic environment as well. The biotic environment consists of other living organisms that affect it—the food it eats, the animals that eat it, the organisms that provide shelter for it, compete with it, pollute its physical environment, cause it to have disease and so on.

The ecology of parasites is unique in that the biotic factors of the environment assume a greater and more direct and continuous role than is true for non-parasitic species. Although physical factors such as temperature have an important impact on parasites, as will be shown in Chapter 8, the intimate relationship between parasite and host is paramount in importance and there is a continual interplay between the biochemistry of the parasite and that of its host. In a real sense the host *is* the environment of the parasite, especially of internal ones. The physical and chemical conditions (temperature, pH, oxygen, etc.) surrounding the parasite are those of the host's body, its food is that of the host or the host's body itself, its shelter is the body of the host, and it depends on the host for transport. Where there are more than one host species in the life cycle, one host may be the means of transferal to another one.

As dependent as parasites are upon the host, the ecological effect is not one way. Parasites are often important parts of the host's biotic environment. Parasitic diseases either alone or in conjunction with other environmental stresses may influence weight or reproduction of the host species and may alter its population characteristics (Chapter 6), and many may thus have great economic importance (Chapter 9).

In this book, three general categories of the ecology of parasites are treated: (1) The ecological effects of the physical environment and of the host upon parasites (Chapter 6). (2) The effects that different species of parasites occupying the same host have on each other (Chapter 7). (3) The ecological effect that parasites have on hosts (Chapters 6 and 9). Ecological as well as historical factors determine the geographical distribution of parasites; this topic is discussed in Chapter 8.

Wherever feasible, examples of *marine* parasites have been used, but in some cases where relevant studies of marine species have not been made, non-marine examples had to be included to elucidate important parasitological principles.

Many groups of animals contain only parasitic species and a non-specialist reader may not be familiar with either the scientific name or common name of such groups. Indeed, in most cases there is no common name. A further complicating factor is that in many species the different developmental stages are quite different in appearance and to the untrained eye would even appear to be different species. Each of these stages must have a name in order to distinguish it from others. In different groups of parasites, such stages are not equivalent in appearance or other characteristics and so different names are applied. Finally, the anatomy of parasites includes a number of structures of ecological importance to the parasite which are not found in non-parasitic animals, especially organs used to attach to the host. The nature of such organs differs among groups of parasites and hence there are a number of anatomical terms unfamiliar to the non-parasitologist.

All these conditions result in a rather bewildering set of terms and names which are necessary even for an introductory, general treatment of the topic, such as set out in the present book. In order to assist the non-specialist reader in understanding the terminology, a number of aids are provided: in Chapter 1, the most important parasitological terms are discussed, and Chapters 2 and 3 contain many examples of marine parasites and of hosts of such parasites including a large number of figures; there is an extensive glossary of terms; the Appendix and Figure 1 set out the classification of all animals so that the reader can readily locate any particular group of parasite or host; finally, Table 1 shows the distribution of parasites in the animal kingdom. Frequent reference to these aids will probably be required by many readers.

The following persons critically read all or large parts of the manuscript and made many useful comments: Dr Lester Cannon, Queensland Museum; Dr David Gibson, British Museum (Nat. Hist.); Professor Harold Heatwole (the series editor) and Mr Russ Hobbs, University of New England; Dr Arlene Jones and Dr C.F.

Khalil, Commonwealth Institute of Helminthology; Dr Bob Lester, University of Queensland.

The following persons commented on small parts of the manuscript: Dr G. Boxshall, Miss A.M. Clark, Dr P. Cornelius, Dr J.D. George, Mr C.G. Hussey, Dr R.W. Sims and Dr J.J. Taylor, all of the British Museum (Nat. Hist.); Dr A.H. McVicar, Marine Laboratory, Aberdeen; and Dr E.R. Noble, University of California. I am grateful for their help.

The following persons and publishers gave permission to reproduce figures: Springer-Verlag, figures 6, 8, 14, 26, 53, 54, 56; W.B. Saunders, Philadelphia, figures 2, 3, 13, 17, 20, 31, 39, 49; *Journal of the Fisheries Research Board of Canada*, figures 4, 5; VEB Gustav Fischer Verlag, figures 7, 9, 36, 43–48, 50, 62, 64; American Society of Parasitology, figures 12, 15; Academic Press, figures 16, 28, 57; Department of Biology, Tulane University, figure 18; Oliver and Boyd, figures 19, 22, 34, 38, 71; McGraw-Hill, figures 21, 29; Australian and New Zealand Association for the Advancement of Science, figures 23–25; Department of Zoology, Victoria University of Wellington, figure 27; Geest and Portig, Leipzig, figure 30; Masson S.A., Paris, figures 32, 33, 35, 37; University of Illinois Press, figures 40–42, 51, 52; Birkhäuser Verlag, Basel, figures 60, 65, 67, 69, 70; University of Miami, figure 66; *Naturwissenschaftliche Rundschau*, Stuttgart, figure 68; Blackwell Scientific Publications, figures 61, 63; Dr Ginetzinskaya, Leningrad, figure 58; Dr Eibl-Eibesfeld, Seewiesen, figure 59; Dr Marshall Laird, St. John's, figures 10, 11; Dr David Gibson, London, figure 55; Dr K. N. Lyons, London, figure 57; Dr Thomas Cheng, Charleston, figures 2, 3, 13, 16, 17, 20, 31, 39, 49; Professor Grell, Tübingen, figures 6, 8, 14; Dr C.J. Sindermann, Miami, figure 28; Professor G. Osche, Freiburg, figures 53, 54; Dr Dogiel, *General Parasitology*, figures 19, 22, 38; Drs Dogiel, Petrushevski and Polyanski, *Parasitology of Fishes*, figures 34, 71. Figs. 40–42, 51, 52 from Baer: Ecology of Animal Parasites, the University of Illinois Press, Urbana, 1952, copyright 1951 by the University of Illinois Press.

Mrs Viola Watt and Mrs Ursula Rohde typed and checked the manuscript. I am grateful for their help. Walter Deas, Sydney, kindly supplied the colour photos for the cover and for figures 5 and 6 on Plate VI.

CHAPTER 1

The Nature of Parasitism

Definition of parasitism; types of parasites and hosts; other associations

There are almost as many definitions of parasitism as there are authors writing about parasites. It is beyond the scope of this book to discuss such definitions at length be they from an ecological, physiological or immunological point of view. *Parasitism*, as used here, is understood to be a close association between two organisms, one of which, the parasite, depends on the other, the host, deriving some benefit from it without necessarily damaging it. A parasite is usually smaller than its host.

A number of different types of parasites can be distinguished. An *obligatory parasite* is a parasite which cannot survive without its host, a *facultative parasite* is a parasite for which a host is optional. *Ectoparasites* live on the surface of the host, *endoparasites* live in its interior. A *permanent parasite* is a parasite which is associated with a host for prolonged periods, a *temporary parasite* lives in or on a host only for short periods. *Larval parasites* are those parasites which are parasitic during one or several larval stages, *adult parasites* live on or in a host during part or the whole of their mature phase. *Periodic parasites* visit their hosts at intervals. A *hyperparasite* is a parasite on or in another parasite, and there may be *hyperparasites* of the first, second, third or even fourth degree. *Intraspecific parasitism* is an association of two individuals of the same species, for instance a male parasitizing a female of its own species. *Latent parasitism* is parasitism which has no obvious effects on the host.

We can also distinguish various types of hosts. There are intermediate, definitive or final, paratenic hosts and vectors. The terminology as used by various authors is not always uniform. In this book the terms are used in the following ways: an *intermediate host* is a host which harbours the developing but sexually immature or larval stage of a parasite, the *definitive or final host* is a host which harbours the sexually mature parasite. *A paratenic or transport host* is a host which serves for dispersing the parasite species, but in which there is no development of the parasite. *Vector* is a very wide term which applies to hosts which carry an infective stage of a parasite (for a detailed account of "hosts" see Odening 1976).

Types of associations which are related to parasitism are phoresis, commensalism, mutualism, symbiosis and predation. Caullery (1952) discussed such associations in the marine environment. In *phoresis* (*phoresy*), one organism uses another as a means of transport without establishing a close association. In *commensalism*, one organism uses food supplied in the external or internal environment of a host (for instance the lumen of its intestine) without affecting it in any way. In *mutualism*, two organisms live together and each of them derives a benefit from the association; however, the association is not compulsory. In *symbiosis*, two organisms derive benefit from the association, and they cannot live without each other. In a wider sense, the term symbiosis is sometimes used to describe all kinds of associations between organisms, including parasitism. In *predation*, one organism, usually the larger one, kills and eats another. However, strict borderlines cannot be drawn between the various types of associations. For instance, a parasite may become a predator by killing its host, or it may be of some benefit to the host species, that is the relationship may become mutualistic. A commensal may affect a host and sometimes damage it and thus become a parasite. For a further account of the terminology used in the study of parasitism, related associations and adaptations to parasitism, the reader is referred to Odening (1974).

Ecological, medical and economic importance of parasites in general

The large number of parasitic species itself indicates how important they are. Arndt (1940) counted 10,000 parasite species out of a total of 40,000 animal species then known in Germany. Although most or at least a very large proportion of the parasites of larger terrestrial and freshwater animals, especially of vertebrates, are probably known in the scientifically advanced countries of Europe and North America, relatively few studies have been made of the parasites of invertebrates. Some exceptions are parasites of freshwater and terrestrial snails and certain important insects. It is certain, therefore, that the proportion of parasitic species is much greater than the 25 per cent given by Arndt. Price (1977), on the basis of a study of the British fauna, estimated that more than 50 per cent of all organisms are parasitic, excluding temporary parasites such as mosquitoes or leeches, but including species intimately associated with plants among the parasites. In the tropics, the proportion of parasitic species may be even higher (Rohde 1976a, 1977b), although this may not apply to all groups of parasites, for example insect parasitoids (Hespenheide 1979).

Any textbook on medical or veterinary parasitology shows how important parasites are as agents of disease in man and domestic animals. It is probable that in spite of advances in the prevention and treatment of parasitic infections, diseases caused by parasites even today are more common than any other kind of disease, particularly in subtropical and tropical countries. Malaria parasites, for instance, affected approximately 350 million people, with 3.5 million deaths annually in the years just following the Second World War, and although the incidence decreased to about 100 million with 1 million deaths in succeeding years, it is again on the increase in many countries. Other parasitic diseases such as trypanosomiasis (caused by flagellates living in the blood and nervous system) and bilharzia (caused by flukes living in the blood of man) also have a wide distribution. Losses of domestic animals due to parasitic infections reach millions every year in Australia alone.

Present state of marine parasitology

There is no reason to believe that marine parasites are ecologically less important than terrestrial and freshwater parasites, and in recent years it has become more and more evident that they are indeed of great ecological, economic and hygienic significance (Schäperclaus 1954, Reichenbach-Klinke and Elkan 1965, Sindermann 1966, 1970, Cheng 1967). Many species of parasites affect marine fish and mammals making them commercially less valuable and probably limit their populations; others lead to mass mortalities, for example in oysters and other molluscs, and still others may be transmitted from marine animals to man. The great number of parasitic species of marine animals which have been described (see Yamaguti 1958–1963b) indicates that parasites play an important part in the ecology of the oceans. The 1,000 fish species occurring in the vicinity of Heron Island at the southern end of the Great Barrier Reef are probably infected with about 2,000 species of monogeneans alone

(Rohde 1976a, 1977b), and the total number of fish parasites in this region was estimated to be in the neighbourhood of 20,000. This compares with less than 150 coral species and less than 120 planktonic species known from the same locality (Sale *et al.* 1976).

In spite of their importance, marine parasites are probably the least known group of organisms. Considering the large number of marine hosts, especially in the tropics, it is no exaggeration to say that the description of marine parasites has hardly begun. Extensive surveys of parasites of marine fishes, not limited to a single parasite group, have been made only in a few northern seas (Osmanov 1940, Shulman and Shulman-Albova 1953, Polyanski 1955, Reshetnikova 1955, Zhukov 1960, Strelkov 1960, Campbell *et al.* 1980).

It is the purpose of this book to draw attention to the wide variety of parasitic life in the sea and to stimulate interest in the study of its role in the ecology of the oceans.

CHAPTER 2

The Kinds of Marine Parasites

Parasites are smaller than their hosts and more difficult to study. We must assume, therefore, that the relative number of parasitic species is. Seven phyla, some of them very large, consist entirely of parasites, available show how widespread and common the parasitic way of life is. Six phyla, some of them very large, consist entirely of parasites, and many other phyla contain a large proportion of parasites. Many of these parasites are marine, as indicated for instance by the lists of marine parasites in Florida compiled by Hutton and Sogandares-Bernal (1960) and Hutton (1964).

Bibliographies on Australian marine parasites can be found in Young (1939), Johnston (1952), Young (1970), Jetté (1977), Roubal (1979), Armitage (1980), Byrnes (1980) and Hooper (1980).

PROTOZOA

Four phyla of single-celled animals are exclusively parasitic and the three others are partly so. Dogiel (1965) estimated that the parasitic Protozoa comprise about 800 marine, 1,000 freshwater and 1,800 terrestrial species, and according to Lom (1970) more than 550 species have been described from marine fishes. However, more recent estimates gave much higher numbers of parasitic species (see Table 1). Records of protozoan parasites of marine fishes in Australia, New Zealand and other South Pacific regions were given by Mackerras and Mackerras (1925) and Laird (1950, 1951a,b, 1952, 1953a,b).

SARCOMASTIGOPHORA

Mastigophora. Among the flagellates, more than sixty species of the dinoflagellate family Blastodinidae are parasites mainly of marine

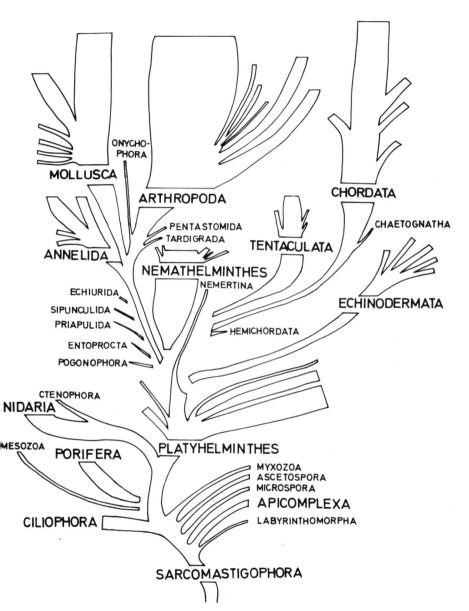

Fig. 1. A phylogeny of animal phyla. Soma parts based on Hanson (1972) but strongly modified. Note: there is no universally accepted genealogical tree of the animal groups.

Table 1. Estimated number of parasitic species among various groups in the animal kingdom. Groups which consist entirely of parasitic species are in italics. Protozoa mainly according to Levine *et al.* (1980), other groups according to various authors, mainly Kaestner (1954–1963), and Barnes (1974, Gnathostomulida). Total number of insect species according to Kaestner, number of parasitic insect species based on the estimates by Price (1977) for the British fauna. Only living species are included.

Phylum	Subphylum (Protozoa) or Class	Total number of species	Number of parasite species
Subkingdom Protozoa			
Sarcomastigophora	Mastigophora	6,900	1,800
	Opalinata	150	150
	Sarcodina	11,500	250
Labyrinthomorpha		some	some
Apicomplexa ⎫			
Microspora ⎬		5,600	5,600
Ascetospora ⎬			
Myxozoa ⎭			
Ciliophora		7,200	2,500
Subkingdom Metazoa			
Mesozoa		55	55
	Dicyemida	40	40
	Orthonectida	15	15
Porifera		5,000	1 (few?)
Cnidaria		8,900	20
	Hydrozoa	2,700	few
	Scyphozoa	200	0 (?)
	Anthozoa	6,000	few
Ctenophora		80	1 (perhaps few)
Platyhelminthes		>8,000	Approx. 6,700
	Turbellaria	>1,600	>80
	Trematoda	4,000	4,000
	Monogenea	1,100	1,100
	Cestoda	1,500	1,500
	Gnathostomulida	80	0
Priapulida		3	0
Entoprocta		60	0
Nemertina		750	10
Nemathelminthes		>12,000	Approx. 5,600
	Rotatoria	1,500	20
	Gastrotricha	150	0
	Nematoda	>10,000	>5,000
	Nematomorpha	<100	<100
	Kinorhyncha	100	0
	Acanthocephala	500	500
Annelida		7,000	Approx. 420
	Polychaeta	4,000	20
	Myzostomida	111	111
	Oligochaeta	2,400	40
	Hirudinea	300	250

Phylum	Subphylum (Protozoa) or Class	Total number of species	Number of parasite species
Onychophora		70	0
Tardigrada		180	1 (?)
Pentastomida		75	75
Arthropoda		800,000	>390,000
	Xiphosura	5	0
	Arachnida	30,000	4,300
	Pycnogonida	350	<350
	Crustacea	20,000	>2,500
	Myriapoda	10,500	0
	Insecta	750,000	>380,000*
Tentaculata		Approx. 5,000	0
	Phoronidea	18	0
	Bryozoa (Ectoprocta)	>4,000	0
	Brachiopoda	280	0
Mollusca		112,000	>100
	Solenogastres	140	few (?)
	Placophora	1,000	0
	Gastropoda	85,000	100
	Scaphopoda	300	0
	Bivalvia	25,000	⩾3
	Cephalopoda	>600	0
Echiurida		70	1 (intraspecific parasite)
Sipunculida		250	0
Hemichordata		80	0
	Enteropneusta	60	0
	Pterobranchia	20	0
Echinodermata		Approx. 6,000	⩾2
	Crinoidea	620	0
	Holothuroidea	1,100	1 (?)
	Echinoidea	860	0
	Asteroidea	1,500	0
	Ophiuroidea	1,900	2
Pogonophora		47	0
Chaetognatha		50	0
Chordata		62,000	few
	Total	>1,060,000	>413,000

*Many authors refer to plant-parasitic insects as herbivorous insects.

invertebrates, sometimes found in their intestine or body cavity, sometimes on their body surface or in planktonic eggs. Others live in radiolarians, and *Oodinium (Amyloodinium) ocellatum* lives on the gills of fishes, where it attaches itself by means of cytoplasmic processes which project at the base of a case covering most of the cell. Large cells drop off the host, the case is closed, cell division occurs and numerous flagellated cells escape, divide once more, infect a host and secrete a new case. It is of special interest that even among endoparasitic species some have not lost their plastids and are capable of photosynthesis. Figure 2 illustrates a species from the gut and coelom of copepods, with the characteristic transverse and longitudinal grooves of dinoflagellates, each containing one flagellum. Figure 3 illustrates a greatly modified species on the surface of a copepod.

Many species of *Trypanosoma* are blood parasites of rays, sharks and teleosts (Fig. 4). Some species are ingested by blood sucking leeches, lose their flagella in the stomach, divide repeatedly and again become flagellated forms which migrate into the proboscis and are ready to infect another fish (Khan 1976). The Australian eels, *Anguilla reinhardtii*, *A. mauritanica* and *A. bengalensis*, are infected with *T. anguillicola* (Johnston and Cleland 1910). Little is known about migrations of Australian eels, and nothing about whether the infection is acquired in the sea or in freshwater. A number of trypanosome species have also been described from New Zealand marine fishes (Mackerras and Mackerras 1925, Laird 1951a, 1952, 1953a). The genus *Hexamita* is a parasite of the intestine of marine fishes, and one species in a marine eel may also occur as a hyperparasite in the eggs and female reproductive system of a trematode living in the eel's intestine. The genus is also found in the intestine of oysters.

Opalinata. Only the genus *Protoopalina* is parasitic in the intestine of marine fish (Fig. 13).

Sarcodina. Few species of parasitic marine amoebas are known. Among them are two species parasitizing arrow worms (Chaetognatha), and others infecting marine algae and eelgrass (*Zostera*).

LABYRINTHOMORPHA

Species of this group are parasitic on algae, mostly in estuarine and marine waters (Levine *et al.* 1980).

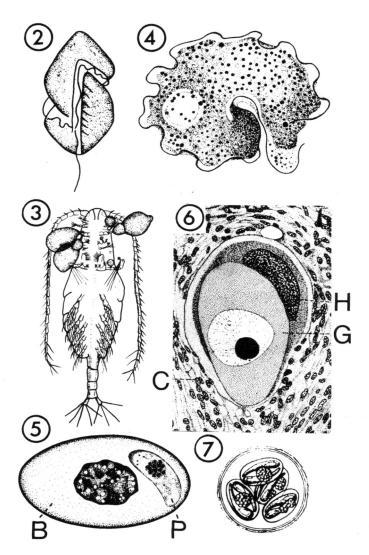

Figs. 2 - 7. Marine parasitic Protozoa.

2. *Syndinium turbo* (Sarcomastigophora, Dinoflagellida) from the gut and coelom of marine copepods (from Cheng 1964, after Chatton).
3. *Ellobiopsis chattoni* (Sarcomastigophora, Dinoflagellida) attached to the copepod, *Calanus finmarchicus* (from Cheng 1964, after Caullery).
4. *Trypanosoma rajae* (Sarcomastigophora, Kinetoplastida) from the blood of *Raja radiata* (after Laird 1969).
5. *Haemogregarina delagei* (Apicomplexa, Coccidia) from *Raja erinacea* (after Laird 1969). P = parasite, B = blood cell.
6. Young gametocyte of *Aggregata octopiana* (Apicomplexa, Coccidia) in a hypertrophied host cell of the intestinal mucosa of *Octopus*; C = connective tissue of host, H = host cell, G = gametocyte (from Grell 1956, after Wurmbach).
7. Oocyst with four spores of *Eimeria clupearum* (Apicomplexa, Coccidia) each spore with two sporozoites (from Doflein and Reichenow 1953, after Wenyon).

APICOMPLEXA

Much more common are various species of Apicomplexa. Of these, the gregarines are parasites living in body cavities, intestines or tissues of invertebrates. Many have anterior hook-like, spine-like or other processes serving for penetration into and adhesion to host cells. The coccidians (Fig. 5) have complex life cycles, similar to those of gregarines and resulting in the production of large numbers of infective cells (Fig. 8). The infective stages or sporozoites are enclosed in spores which are formed in oocysts (Fig. 7). A common species is *Eimeria sardinae* which infects the testes of herring and related fish in the Atlantic and adjoining seas and may cause sterility. *Plasmodium*, species of which cause malaria in man, has been reported from penguins (Huff and Shiroishi 1962). Marine invertebrates such as molluscs and polychaetes are parasitized by several genera of coccidians (Fig. 6). *Aggregata eberthi*, for example, parasitizes marine crabs and cephalopods (Fig. 8). Infection of the cephalopod occurs when it eats infected crabs; crabs become infected by eating pieces of the intestinal epithelium of the cephalopod containing parasitic spores. Lobsters have similar parasites possibly belonging to a different genus. Among the coccidians are also blood parasites of the genus *Haemogregarina* (Fig. 5). They live in the blood cells of many marine fish and are probably transmitted by leeches (Laird 1953a, Becker 1970, Kirmse 1978).

MICROSPORA

The microsporans are intracellular parasites infecting many tissues and organs. Like the Myxozoa, their spores have polar capsules containing a filament. The genus *Glugea* is widespread among marine fishes. *G. stephani*, for instance, infects the intestinal wall of various flatfishes. Large cysts containing huge numbers of spores are surrounded by thick walls of connective tissue fibres. Other genera of Microspora commonly infecting marine fishes are *Pleistophora* and *Nosema*. Several species infect marine crustaceans and some are hyperparasitic. An example of a microsporan hyperparasite is *Nosema legeri* in the parenchyma of the adult trematode *Brachycoelum*, and in the metacercaria of *Gymnophallus somateriae*, both parasitic in marine bivalves. Another species, *N. spelotremae*, was found in metacercariae of *Spelotrema carcini* parasitizing the marine crab *Carcinus maenas*.

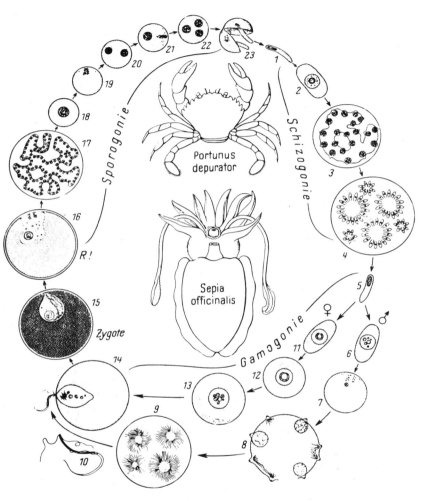

Fig. 8. Life cycle of *Aggregata eberthi* (Apicomplexa, Coccidia) 1 - 4 schizogony in crabs of several species, schizonts in connective tissue cells around intestine; if crab is eaten by *Sepia*, merozoites penetrate into submucosal cells and grow up to gametocytes. 6 - 10 formation of microgametes; 11 - 14 formation of macrogametes; 15 - 23 sporogony: formation of round spores with sporozoites. Infection by means of spores (from Grell 1956, after Dobell).

ASCETOSPORA

The ascetosporans are similar to the microsporans, but their spores do not contain polar filaments. Species of *Haplosporidium* are widespread in marine annelids, but they also occur in nemerteans, gastropods and ascidians. *Minchinia nelsoni* and *M. costalis* are important parasites of oysters, responsible for several mass mortalities and large economic losses (Fig. 16). Sporulation of *M. nelsoni* was recently described by Couch *et al.* (1966).

MYXOZOA

Species of this group are widespread among marine and freshwater fishes and some of them are of great economic importance. Many internal organs and tissues may be infected, for instance the gall and urinary bladders, the kidneys, muscles and nervous system. *Ceratomyxa*, *Myxobolus*, *Myxidium* (Fig. 15), *Chloromyxum* and *Kudoa* (Paperna and Zwerner 1974) are examples. The genus *Kudoa* is of special economic importance (Plate I). Noble (1966) recently described some species from deep water fishes.

CILIOPHORA

Many ciliates are probably not genuine parasites but commensals which use a host only as a substratum for attachment. *Hypocomella cardii* (Fig. 9) is a genuine parasite, deeply anchored by means of long processes in the marine mussel, *Cerastoderma edule*. *Orchitophrya stellarum* parasitizes the testes of the starfish *Asterias rubens* at Plymouth, and causes complete breakdown of the germinal tissue and castration. *Caliperia brevipes* forms a much looser association with its host, the ray, *Raja erinacea*, to whose gills it is attached by means of a loop-like basal appendage (Fig. 10). A common genus on the skin and gills and in various organs of freshwater and marine animals is *Trichodina* (Fig. 11) (Lom and Laird 1969). Echinoderms, molluscs and particularly fish are marine hosts. The parasites attach themselves with a basal sucking disk and if infections are heavy may cause extensive damage to the hosts. *Cryptocaryon irritans* is important in marine aquaria, where it causes a form of white spot disease in fish (see Sindermann 1970, further references therein). A ciliate parasitic on another marine ciliate is *Tachyblaston ephelotensis* (Fig. 14). Its infective stage penetrates the pellicle of the host,

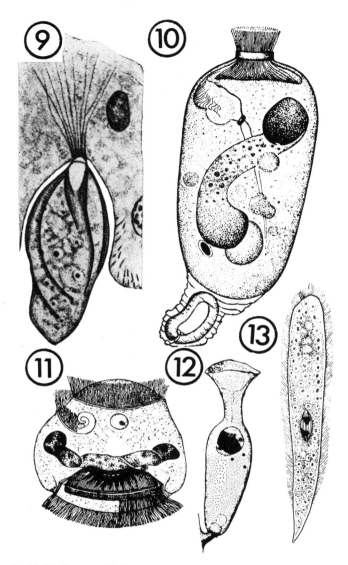

Figs. 9 - 13. Marine parasitic Protozoa.

9. Section through *Hypocomella cardii* (Ciliophora) attached with its processes to host tissue, on gills of *Cerastoderma edule* (from Doflein and Reichenow 1953, after Chatton and Lwoff).
10. *Caliperia brevipes* (Ciliophora) from the gills of *Raja erinacea* (from Noble and Noble 1976, after Laird).
11. *Trichodina parabranchicola* (Ciliophora) from the gills of intertidal fish (from Noble and Noble 1976, after Laird).
12. *Oenophorachona ectenolaemus* (Ciliophora) from Noble and Noble 1976, after Matsudo and Mohr).
13. *Protoopalina saturnalis* (Sarcomastigophora, Opalinata) from large intestine of marine fish, *Box boops* (from Cheng 1964, after Léger and Duboscq).

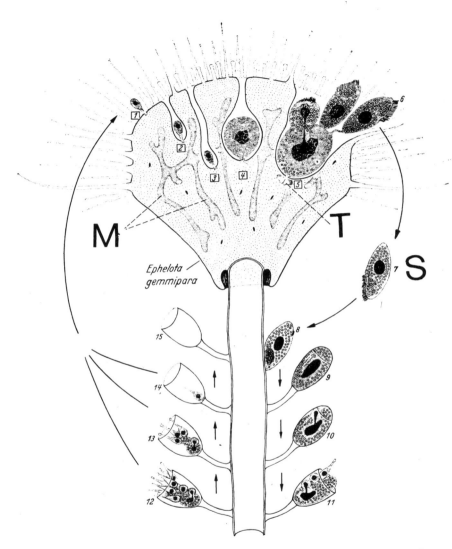

Fig. 14. Development of *Tachyblaston ephelotensis* (Ciliophora, Suctoria). 1. Infective stage (dactylozoite) penetrates pellicle of suctorian *Ephelota gemmipara.* 2 - 4. Stages of penetration and growth. 5. Formation of ciliated swarmers. 6 - 8. Escape, free swimming phase and settling of swarmer. 9 - 15. Formation of stalks and repeated budding; infection of new host (from Grell 1956, after Grell) M = macronucleus of host cell, S = swarmer, T = tentacle of parasite cell.

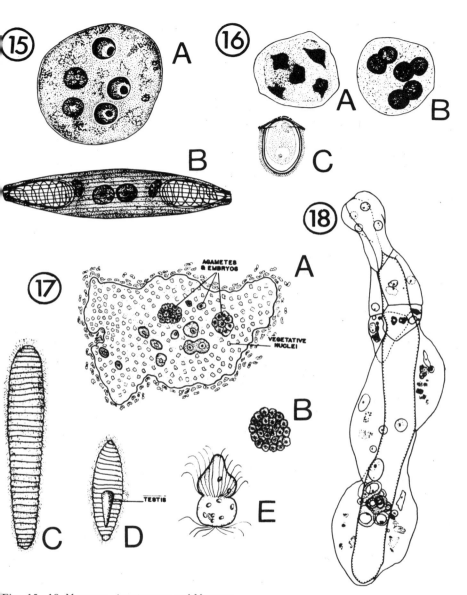

Figs. 15 - 18. Myxozoa, Ascetospora and Mesozoa.

15. *Myxidium coryphaenoidium* (Myxozoa) from *Coryphaenoides* sp. A. Trophic stage. B. Spore (after Noble 1966).
16. *Minchinia nelsoni* (Ascetospora) in *Crassostrea virginica*. A and B. Plasmodia. C. Spore (from Cheng 1967, after several authors).
17. *Rhopalura* sp. (Mesozoa, Orthonectida) A. Male plasmodium from a brittle star. B. Morula. C. Free-living female. D. Free-living male. E. Ciliated larva (from Cheng 1964, after several authors).
18. Nematogen of *Dicyema hypercephalum* (Mesozoa, Dicyemida) from *Octopus joubini* (after Short 1962).

Ephelota gemmipara, with its tentacles, and grows up at the base of a deep invagination of the pellicle. Host cytoplasm is absorbed by these tentacles, and many swarmers are formed which leave the host and attach themselves to a substratum. There they grow up to a stage which is stalked and has an operculate housing. Approximately sixteen infective stages are formed by budding; they escape after the operculum has opened, and infect new hosts (Grell 1956).

MESOZOA

It is not certain that the two groups Dicyemida and Orthonectida, comprising the Mesozoa, are in fact related. The Dicyemida are small parasites (less than 2 mm in length) in the nephridia of cephalopods (Fig. 18). Ciliated larvae leave the host but it is not known how infection occurs. The Orthonectida parasitize various marine invertebrates (echinoderms, turbellarians, molluscs, polychaetes and nemerteans). The species *Rhopalura ophiocomae* (Fig. 17), for example, lives in Ophiuroidea. Male and female parasitic plasmodia produce agametes which become sexual forms. The latter escape from the host, copulate and male and female ciliated larvae are produced which infect a host via the genital pore. They disintegrate and release germ cells which enter host cells and grow up to plasmodia. Only a few species of mesozoans are known and they are without any economic importance.

PORIFERA (Sponges)

A few species of the sponge *Cliona* bore into the shells of marine molluscs (see for instance, Hopkins 1956a, b, 1962). They are a common pest in oyster beds, but they seem to be predators rather than true parasites. Studies with the electron and scanning electron microscopes have shown that *Cliona lampa* burrows into the bivalve shells by means of cytoplasmic extensions and etching secretions of cells. Substratum chips are expelled through the system of exhalant canals (Rützler and Rieger 1973). The sponge, *Iophon*, is a common parasite on the Southern Ocean ophiuroid, *Ophiolepis gelida* (see Fell 1961).

CNIDARIA

Both Hydrozoa and Anthozoa include a few parasitic species and, according to Rees (1967), all major groups of Cnidaria have species which live in a commensal, mutualistic or parasitic relationship with molluscs. The early developmental stages of the hydrozoan, *Polypodium hydriforme*, first described in detail by Lipin (1911), infect the eggs within the ovary of the sturgeon, *Acipenser ruthenus* (Fig. 19). The parasitic stage is inside out and develops a long sac-like stolon from which buds are formed. The stolon bursts after the eggs are laid, and releases polyps which evaginate and become free living. The planula larvae of some other hydrozoans, *Cunoctantha octonaria*, *Cunina proboscidea* and *C. peregrina*, are parasites of the medusae, *Turritopsis*, *Geryonia*, *Rhopalonema*, and give rise to their own medusae by budding. The colonial *Hydrichthys* hydroid lives and feeds on fish, and medusae of another hydrozoan, *Mnestra*, parasitize the marine snail, *Phyllirrhoe*. Conversely, it may be that the snail is the parasite of the medusa (Martin and Brinkmann 1963: snail, *Phyllirrhoe bucephala*, on medusa, *Zanclea costata*). Another hydroid, *Hydractinia echinata*, usually lives on the external surface of the shell of marine snails and mussels. Sometimes it becomes established on the internal surface, where it interferes with the mantle, often causing shell deformities (Merrill 1967). The relationship, therefore, may be considered a parasitic one. Among the Anthozoa, larvae of *Edwardsia* and *Peachia* infect the body surface or gastrovascular cavity of other sea anemones or ctenophores.

CTENOPHORA

Beside *Coeloplana* spp., which live on the surface of certain soft corals and are probably not true parasites, the young of one species, *Gastrodes parasiticum*, live in the mantle of the tunicate, *Salpa fusiformis*. Maturation occurs in the free environment and infection is by planula larvae (Fig. 20). A detailed description was given by Komai (1922).

PLATYHELMINTHES (Flatworms)

Turbellaria. Bresslau (1928–1933) reviewed the parasitic Turbellaria and their hosts. Among the Acoela, *Ectocotyla paguri* attaches itself with a posterior sucker to hermit crabs (Fig. 21) and other

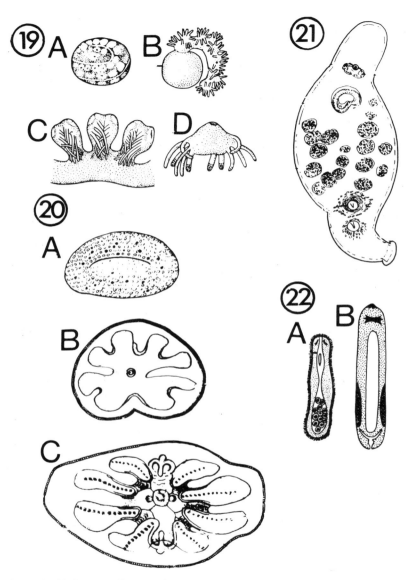

Figs. 19 - 22. Parasitic Cnidaria, Ctenophora and Turbellaria.

19. *Polypodium hydriforme* (Cnidaria). A. Egg of sturgeon with round stolon of *Polypodium*. B. Stolon emerging from egg. C. Part of stolon with three buds turned inside out. D. Free polyp (from Dogiel 1964, after several authors).

20. *Gastrodes* sp. (Ctenophora). A. Bowl-shaped stage parasitic in *Salpa*. B. Later parasitic stage. C. Larva ready to leave host (from Cheng 1964, after Komai).

21. *Ectocotyla paguri* (Turbellaria) from surface of hermit crab (from Hyman 1951, after Reinhard).

22. *Fecampia xanthocephala* (Turbellaria). A. Young form with eyes and mouth opening. B. Adult without eyes and mouth opening (after Dogiel 1964).

species live in the intestine of echinoderms. Various species of Rhabdocoela inhabit marine molluscs, echinoderms, sipunculoids and myzostomids. *Fecampia* matures in the haemocoel of marine crustaceans, losing its eyes, mouth and pharynx and retaining only the intestine. The parasite leaves the host, secretes a capsule containing egg and yolk cells, and dies. Ciliated young worms with eyes and a fully developed digestive tract infect a new host (Fig. 22). According to Christensen and Kanneworff (1976) the Fecampiida have a worldwide distribution and considerably more than the few known species remain to be described.

Females of *Kronborgia caridicola* live in the body cavity of shrimps, where they occupy all the available space when approaching the mature stage. The mature female leaves the host and deposits more than half a million egg capsules, each with two egg cells and many yolk cells, in a cocoon which is attached to the bottom substratum (Kanneworff and Christensen 1966). Among the Tricladida, species of the genus *Bdelloura* have a posterior sucker used for attachment to the horseshoe crab, *Limulus*; other triclads live on skates. However, it is doubtful that these are genuine parasites.

Trematoda. More than 4,000 species of trematodes have been described, and many of these are endoparasites of marine fishes. The small group Aspidogastrea is exceptional in having a simple life cycle and a large ventral adhesive disk with numerous alveoli or a row of suckers. Individuals mature either in snails or bivalves, but also survive when eaten by such vertebrates as fish or turtles. In some species, young stages in molluscs mature only after ingestion by a vertebrate. *Lobatostoma manteri*, for instance, grows up to full body size in the digestive system of several species of marine snails on the Great Barrier Reef, but matures only when eaten with the snail by the fish, *Trachinotus blochi*. Infection of the snails occurs when they eat eggs containing the infective larval stage (Fig. 24; Rohde 1973). The much larger trematode group Digenea contains forms which typically have one anterior and one ventral sucker, one or both of which are sometimes absent. The life cycle involves at least one intermediate host, usually a snail, and several larval stages (Fig. 23). Most Digenea are endoparasitic in the digestive tract and in various other organs of vertebrates, but several species have secondarily adopted an ectoparasitic way of life. *Transversotrema* lives under the scales of marine fishes, for instance tuskfish (*Choerodon*) on the Great Barrier Reef, and *Syncoelium* attaches itself by a long stalked ventral sucker to the spines usually of the gill arches of a great variety of marine fishes in the Pacific Ocean. Marine invertebrates and vertebrates often harbour the metacercarial stage of trematodes, but

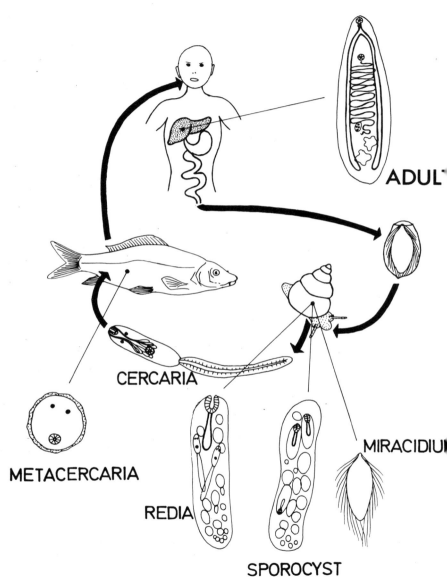

Fig. 23. Typical life cycle of a digenetic trematode (after Rohde 1976c).

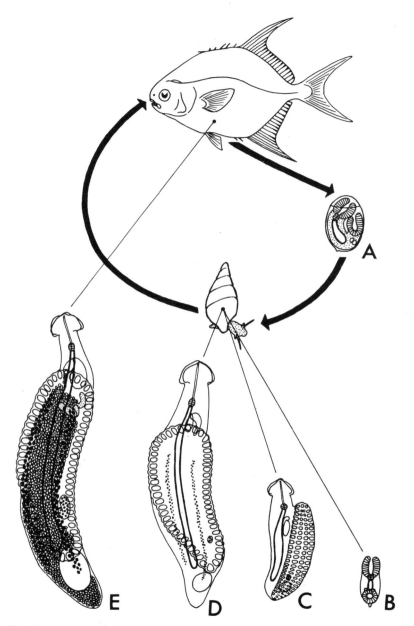

Fig. 24. Life cycle of *Lobatostoma manteri*, Aspidogastrea (after Rohde 1976c). A = egg, B. = larva, C = juvenile, D = pre-adult, E = adult.

these stages may sometimes be progenetic, that is have fully developed reproductive organs and produce sperm and eggs. For instance, sexually mature forms of *Derogenes varicus* have been found in the cephalopod, *Sepia officinalis*, in the copepod, *Lernaeocera lusci*, in the hermit crab, *Pagurus pubescens*, and in the chaetognath, *Sagitta bipunctata* (see Overstreet and Hochberg 1975), and adults of *Proctoeces subtenuis* occur in lamellibranchs (Freeman 1958, White 1972).

Monogenea. Most monogeneans are ectoparasites of fish, and frequently of marine fish. They commonly live on the gills, skin or fins, but some have invaded the rectal cavity and its vicinity in sharks and chimaeras, and others the ureters, body cavities and even the blood system. Attachment is by hooks, clamps, suckers, friction pads, surface spines, cement glands, or a combination of these. There are two groups of Monogenea, the Monopisthocotylea and Polyopisthocotylea. Monopisthocotyleans have a posterior attachment organ (opisthaptor) consisting of a sucker and/or hooks (Fig. 25), whereas the polyopisthocotyleans have an opisthaptor bearing complex clamps and sometimes also hooks (Fig. 26). Unlike the Digenea, the Monogenea have a direct life cycle, that is they do not have an intermediate host. Their larvae hatch from eggs and are gradually transformed into the adult stage. However, it has been suggested that certain marine genera, *Pricea* and *Gotocotyla*, infect gills of various fish where they do not develop beyond a certain stage, and maturation is thought to occur only after these fish have been eaten by the "final" host, large mackerel (fam. Cybiidae) (see page 62). Both these genera are common on the Spanish mackerel, *Scomberomorus commerson*, on the Australian northeast coast.

Cestoda. One group usually included in the cestodes, the Cestodaria, is sometimes considered to be a group of its own. In contrast to most other cestodes, the cestodarians are unsegmented and they have a larva with ten instead of six hooks. The main genera of Cestodaria are *Amphilina* parasitic in the coelom of fishes, and *Gyrocotyle* parasitic in the intestine of chimaeras (Fig. 27). Intermediate hosts of *Amphilina foliacea* are freshwater amphipods, but the final hosts, sturgeons, migrate into the sea. The egg of *Amphilina* contains a ciliated larva which hatches after it has been eaten by an amphipod. The ciliated epidermis is shed in the intestine of the intermediate host, the larva penetrates into its body cavity and develops to the infective stage, the plerocercoid. Fish become infected by eating amphipods containing the larvae. *Gyrocotyle* and related genera are characterized by a posterior funnel-like structure with folded or simple margin

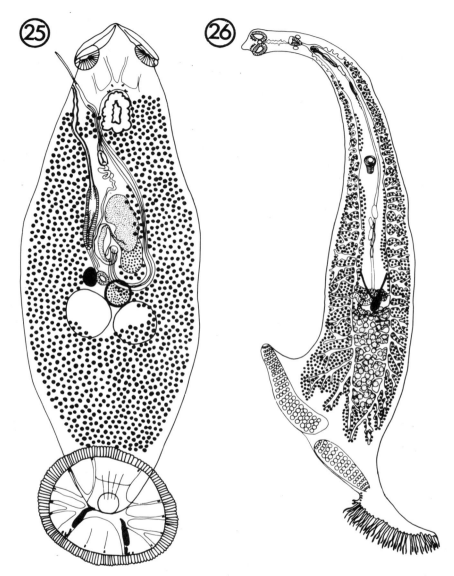

Figs. 25 - 26. Adult Monogenea.

25. *Sprostonia longiphallus* (Monopisthocotylea) from *Epinephelus tauvina* (after Rohde 1976c).
26. *Heteromicrocotyla australiensis* (Polyopisthocotylea) from *Carangoides emburyi* (after Rohde 1977a).

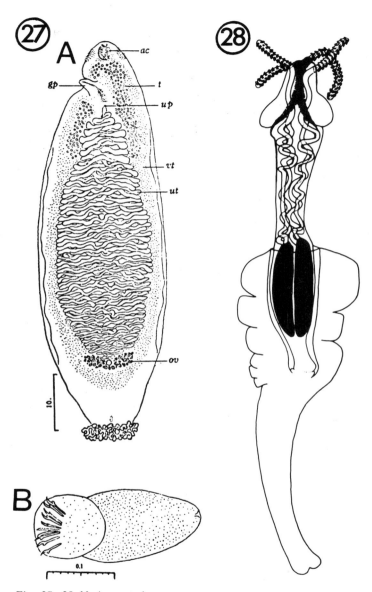

Figs. 27 - 28. Marine cestodes.
27. *Gyrocotyle* (Cestodaria) from *Callorhynchus milii* in New Zealand. A. Adult. B. Larva (after Manter 1951). ac = acetabulum, gp — gonopore, ov = ovary, t = testes, up = uterine pore, ut = uterus, vt = vitellaria.
28. Larval trypanorhynch (Eucestoda) scolex retracted (from Sindermann 1970, after Kahl).

and a dorsal opening, the rosette. It is not known whether intermediate hosts are necessary and how infection of fish occurs. Most cestodes, sometimes referred to as Eucestoda or true tapeworms, have a scolex (head) and a strobila or chain of segments. The scolex usually bears holdfast organs, such as suckers and hooks, and the segments contain the reproductive organs in various stages of development. Occasionally, only a single segment is present. Almost all cestodes have an indirect life cycle with at least one intermediate host. Australian sharks are commonly infected with various species of Trypanorhyncha, a group of tapeworms characterized by four spiny "tentacles" which can be everted (Fig. 28). There are several types of life cycles in this group. In some species eggs are laid and a ciliated six-hooked larva, the coracidium, escapes and develops into the procercoid when ingested by a copepod, and the procercoid becomes a plerocercoid in the second intermediate host, a plankton eating fish; the life cycle is completed in a shark which has eaten an infected fish. In other species, filter-feeding molluscs or two subsequent teleost fish serve as intermediate hosts.

NEMERTINA

Several species of Nemertina are either parasites or commensals of marine invertebrates. They have been found associated with crustaceans, sea anemones, tunicates and clams. The species *Malacobdella grossa* in the European bivalve, *Zirfaea crispata*, was studied in detail by Gibson (1967, 1968) and Gibson and Jennings (1969). They found that there were no measurable effects on the host and that the diet consisted of free-living organisms. Thus the species is a commensal rather than a parasite. Hyman (1951) also stated that there are no true parasites among the nemertines, although some show changes characteristic of parasites, for instance reduction of certain sense organs and increase in reproductive capacity (*Gononemertes*, Fig. 29). However, Humes (1942) and Hopkins (1947) showed that *Carcinonemertes carcinophila* feeds for certain periods on the eggs of its crab host and thus may be a true parasite.

NEMATHELMINTHES

Rotatoria (Rotifera). Only a few rotifers are associated with marine hosts. Several species of the genus *Albertia* live in the intestine (and in

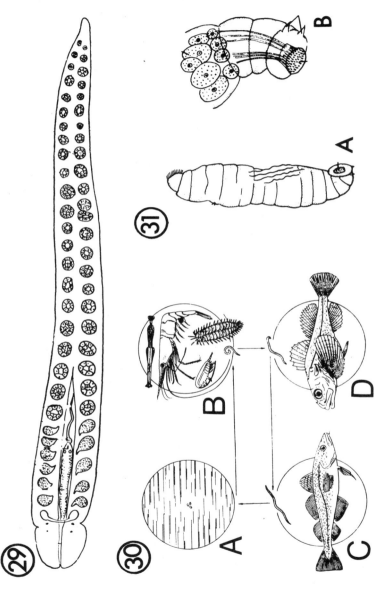

Figs. 29 - 31. Parasitic Nemertina, Rotatoria and Nematoda.
29. *Gononemertes* (Nemertina) (from Hyman, 1951 after Brinkmann).
30. Life cycle of *Thynnascaris aduncum* (Nematoda). A. Egg in water. B. Second stage larva in first intermediate hosts. C. Third stage larva in second intermediate host. D. Adults in final hosts (from Odening 1969, after Uspenskaja).

one case also on the surface) of marine annelids, and *Proales gonothyraeae* in the theca of an hydroid. *Zelinkiella synaptae* (Fig. 31) attaches itself with its posterior adhesive disk to the surface of sea cucumbers, and species of the genus *Seison* have a similar disk used for attachment to marine crustaceans. It is possible that the relationship is commensal rather than parasitic, although the latter species may sometimes feed on the host's eggs. That rotifers can form intimate associations with their hosts, far beyond mere commensalism, has been shown by histological studies of *Albertia vermiculus* in the intestine of earthworms (Rees 1960). Thane-Fenchel (1966) gave evidence that the rotifer, *Proales paguri*, grazes on the epithelial cells of the gills of the hermit crab, *Pagurus bernhardus*.

Nematoda (Roundworms). One of the most uniform and at the same time one of the largest groups in the animal kingdom is the Nematoda, many of which are parasites of marine animals. Nematode parasites of marine fish were discussed by Margolis (1970). A typical roundworm has an elongate cylindrical shape with a straight digestive tract, and the sexes are separate. The male is usually smaller than the female. Four moults during development result in a life cycle comprising four larval stages and the adult. Infection occurs usually at the third larval stage. Intermediate hosts may be present or absent. In species of *Thynnascaris* for example, various invertebrates serve as first hosts and various fishes as second intermediate hosts; final hosts are large fish which feed on the second intermediate hosts (Fig. 30). In *Phocanema* first, second and final hosts, respectively, are crustaceans, fishes and seals, and *Contracaecum* has mammals and birds as final hosts. Invertebrates may be hosts to adult nematodes, for instance the fiddler crab, *Uca mani*, is parasitized by *Rhabdochona uca*, but they serve more commonly as intermediate hosts. Thus, lobsters are frequently, and oysters sometimes, infected with larval roundworms which mature in fish. Hookworms infect young seals (Olsen and Lyons 1965), and there are probably few marine vertebrates which are not host to at least one species of nematode.

Nematomorpha (Horsehair worms). Most species of the horsehair worms superficially resemble nematodes, but they are even longer and the adults have a non-functional intestine. Nearly all species live in terrestrial or freshwater habitats, and only one genus, *Nectonema*, is marine (Fig. 32). The larva of *N. agile* parasitizes marine crustaceans (Fig. 33).

Acanthocephala (Thorny-headed worms). The thorny-headed worms can easily be recognized by their retractible proboscis bearing many thorns or hooks. The adults are parasites in the digestive tract of vertebrates, particularly of fishes (review by Crompton 1970). The

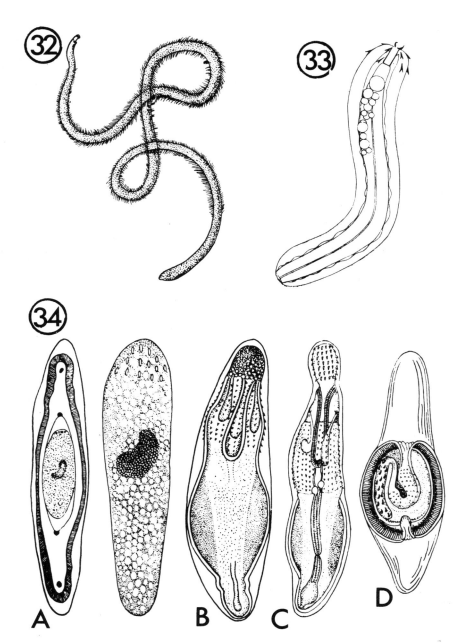

Figs. 32 - 34. Nematomorpha and Acanthocephala.

32. *Nectonema* (Nematomorpha), adult (from Grassé 1965, after Feyel).
33. *Nectonema* (Nematomorpha), larva (from Grassé 1965, after Huus).
34. Larval Acanthocephala. A. Egg. B. Acanthor. C. Acanthella. D. Cystacanth (from Dogiel *et al* 1961, after Petrochenko).

female worms lay eggs which are eaten by benthic crustaceans, in which the first larva, the "acanthor", hatches, develops to the "acanthella" and further to a "juvenile". The juvenile may encyst thus becoming a "cystacanth" (Fig. 34). The vertebrate final host usually becomes infected by ingesting infected crustaceans. Sometimes other invertebrates (snails) or fish may serve as transport hosts, in which the infective larva survives and retains its infectivity, without undergoing further development. Examples of acanthocephalans with marine fish as transport hosts are several species of *Corynosoma*, which mature in seals and sea birds. According to Crompton (1970), the "transport host" of certain species may in fact be a necessary second intermediate host in which growth occurs.

ANNELIDA (Segmented worms)

Polychaeta. Only a few species of the almost exclusively marine polychaetes have adapted to a parasitic way of life. *Ichthyotomus sanguinarius* lives on fish, especially eels, in the Mediterranean. It feeds on the hosts' blood sucked in by means of an anterior sucker-like structure (Fig. 37). *Parasitosyllis* parasitizes nemerteans and polychaetes, *Histriobdella* lives in the branchial chamber of lobsters, and several other species are parasitic in the body cavity and vascular systems of polychaetes. Many other species live in more or less loose associations with various hosts (Clark 1956). The "mudworms", *Polydora* spp., for instance, live in mud filled burrows in the shells of oysters and other bivalves. The species *P. ciliata* has been held responsible for mass mortalities of oysters in southeastern Australia (Whitelegge 1890, Roughley 1922, 1925).
Myzostomida. The myzostomids are ecto- or endoparasites of echinoderms and particularly of crinoids. They are, besides parasitic snails, the oldest fossil parasites known; galls probably produced by them have been found on the arms of feather stars from the Silurian and Devonian. Typically, a myzostomid is dorso-ventrally flattened with five pairs of short parapodia and a branched intestine (Fig. 36). Among the species creeping around on the surface of their hosts, many are probably not true parasites but commensals which feed on detritus, others bore into the host and are often enclosed in a gall-like structure. *Protomyzostomum polynephris* is a parasite in the coelom of an ophiurid, and some species live in the intestine and coelom of crinoids.
Hirudinea. No marine oligochaete parasites are known, but a number of leeches are ectoparasites of marine animals, feeding on

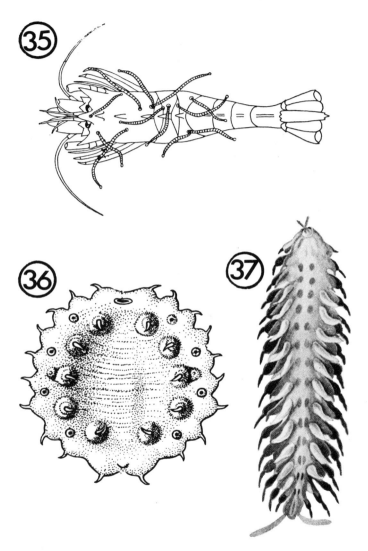

Figs. 35 - 37. Parasitic polychaetes, Myzostomida and Hirudinea.

35. *Crangonobdella murmanica* (Hirudinea) on *Sclerocrangon boreas* (from Grassé 1959, after Selensky).
36. *Myzostoma* sp. (Myzostomida), ventral view (from Kaestner 1954).
37. *Ichthyotomus sanguinarius* (Polychaeta) attached to the fin of *Myrus* (from Grassé 1959, after Eisig).

their blood. Contrary to the other annelids, usually they do not have bristles, and they have an anterior and a posterior sucker. Species of the genus *Pontobdella* parasitizing elasmobranchs, and *Hemibdella soleae* parasitizing the European sole are examples. Figure 35 shows a number of leeches attached to a marine crustacean.

ARTHROPODA

Arachnida. Mites and ticks parasitize sea birds. Otherwise, among the arachnids, only a few mites are marine parasites. *Orthohalarachne* spp. and *Halarachne* spp. live in the respiratory passages of seals. The body of the female *Halarachne* is long and worm-like; it bores into the mucosa and is anchored by means of the claws of the anterior legs. Destroyed tissue and lymph are ingested. Males are rare and the larvae can move freely on the body surface and are apparently responsible for the infection of other hosts. Another species, *Enterohalacarus minutipalpis*, lives in the intestine of the sea urchin, *Plesiodiadema indicum*.
Pycnogonida. The sea spiders are exclusively marine. Some species are adult and apparently most species are larval parasites. Hosts of the larvae are mainly Cnidaria, but also Ectoprocta, Ascidia, Echinodermata, Mollusca, Porifera, Brachiopoda, Annelida and in some cases even brown and red algae. Some are ectoparasites but others are endoparasites (Fig. 38). In ectoparasitic species, the larvae pierce the surface of their host and feed on tissues and body fluids; in some intermediate forms the proboscis penetrates through the mouth opening into the gut of the host sucking in the contents of the gut as food, and in certain endoparasitic species the larvae are surrounded by gall-like growths of the host. In a typical life cycle, eggs laid by the female are carried in clusters by specially adapted legs of the male, the so-called ovigers. Larvae hatch and are sometimes found for a long time on the male. They are either gradually transformed into the adult or pass through a special four legged larval stage, the protonymphon. Adults of most species are predatory on soft animals, often polyps of colonial hydrozoans, but they may also be considered ectoparasites of hydrozoan colonies, because only parts of the colonies are usually eaten; the adults of some species are ectoparasites of larger cnidarians (Fig. 39).
Crustacea. The copepods represent one of the largest groups of ectoparasites of marine fish (Mann 1970) and other marine animals. *Cymbasoma rigidum*, for instance, is a larval parasite of polychaetes and probably of molluscs. The nauplius larva attaches itself with its

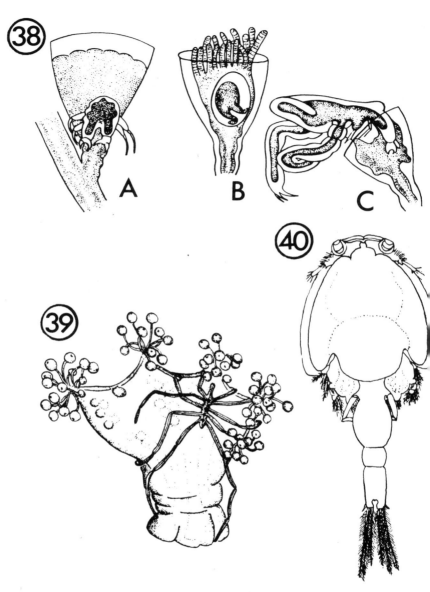

Figs. 38 - 40. Pycnogonida and parasitic crustaceans.

38. Larval pycnogonids. A. *Ammothea laevis* at base of a hydranth of *Obelia*. B. *Anoplodactylus petiolatus* in gastral cavity of a hydranth of *Obelia*. C. *Endeis spinosus*, feeding on hydranth of *Obelia* (after Dogiel 1964).

39. Ectoparasitic young adult of *Phoxichilidium femoratum* (Pycnogonida) on the stauromedusa *Lucernaria* (from Cheng 1964, after Helfer and Schlottke).

40. *Caligus rapax*, ♂ (Copepoda) (after Baer 1952).

mandibles to the host, penetrates into a blood vessel and forms two long processes which absorb food. A later stage escapes from the host, moults and begins a brief pelagic life. Species of the families Notodelphyidae, Namakosiramiidae, Myzopontiidae and Ascomyzontidae are parasites of tunicates, echinoderms, lobsters and other marine animals, and numerous species of the genus *Caligus* and of related genera (order Siphonostomatoida) are among the most common ectoparasites of fish (Fig. 40). The early larval stages are free living and for certain species it has been shown that the adults may also leave the host, live in the plankton and infect another host. There are many parasitic species in the families Penellidae (*Lernaeocera*, *Penella*) and Lernaeopodidae. Most live on fishes, but some (*Penella*) also on cephalopods and whales. *Lernaeocera branchialis* has economic importance; it infects the gills of many marine fish, and penetrates into the host's blood system. A list of British invertebrates with their known parasitic and commensal copepods was given by Leigh-Sharpe (1935–36), and a recent detailed account of British parasitic copepods is by Kabata (1979).

The Branchiura are temporary parasites which visit a fish only to suck blood. *Argulus*, and other genera, include marine species (Mann 1970) (Fig. 41).

In the Cirripedia (barnacles), all 25 species of Ascothoracida are parasitic, with one possible exception. *Ascothorax ophioctenis*, which lives in the bursae of an ophiuroid, *Laura*, which is parasitic in gorgonians and antipatharians, and *Dendrogaster* (Fig. 42) in the coelom of sea stars are examples. The latter genus shows strong morphological reductions due to the parasitic way of life. The thorax and abdomen are indistinctly segmented, and the mantle is strongly branched. The males are very small, living in a cavity formed by the female.

Also among the Cirripedia, many species of Thoracica have adopted a parasitic way of life. *Anelasma*, a genus related to the goose barnacle *Lepas*, parasitizes the skin of a shark, and processes of its stalk branch in the host's musculature. All the approximately 120 species of rhizocephalans are parasitic. They form reticula of tissue in their hosts, only the brood sac with the gonads appearing on the surface. The adult has lost all the characteristics of cirripedians and crustaceans and only the larvae indicate its true nature. A common genus is *Sacculina* in crabs (Fig. 43). Its early larval stages are free living, the cypris larva attaches itself to a host, a cell cluster migrates into the host's thorax and begins to grow processes around the intestine, muscles, nerves and gonads of the host. Finally, a reproductive sac appears on the surface through a hole in the shell of

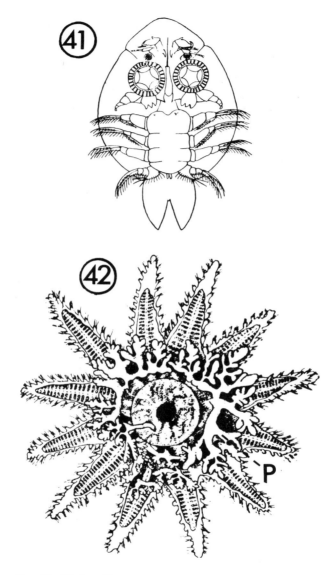

Figs. 41 - 42. Parasitic crustaceans.
41. *Argulus* (Branchiura) (from Baer 1952, after Wilson).
42. *Dendrogaster murmanensis* (Ascothoracida) in a sea star (from Baer 1952, after Korschelt). P = parasite.

the crab. Appendages, most sense receptors, intestine and excretory organs of the parasite have disappeared as a consequence of the endoparasitic way of life.

It seems doubtful that there are true parasites among the Decapoda, although many live in a commensal association with corals, mussels, ascidians, etc. Small crabs of the family Pinnotheridae, for instance, are such commensals.

Many parasites are found among the Isopoda. The Gnathiidea (*Gnathia*, etc.) are free living and do not feed in the adult stage. They use up blood ingested by the parasitic larval stage which parasitizes fish (Fig. 44). Gnathiids are common on Australian marine fishes and can easily be recognized by the swollen thorax due to the volume of blood ingested. A key to seventeen Australian species was given by Holdich and Harrison (1980). Among the Anthuridea, the genus *Paranthura* parasitizes fish, and many adult Cymothoidae are fish parasites, occasionally also occurring on cephalopods. Species of the family Bopyridae are parasites on decapod crustaceans. The males are very small and the female is often strongly asymmetrical. An example is *Cancricepon* (Fig. 45). Dajidae are parasitic on Mysidacea and Eucarida, and the Entoniscidae in the branchial and visceral cavities of crabs and other decapods (Hartnoll 1960). The female entoniscids have strongly deformed bodies (Fig. 46) and the males are very small; larvae infect pelagic copepods and are transformed into juveniles which search for the final host, at least in the only species studied, *Portunion kossmanni*. The Cryptoniscina infect various groups of crustaceans, for example ostracods, barnacles, amphipods and isopods. *Danalia* is a hyperparasite of the rhizocephalan, *Sacculina* (Fig. 47).

Less common as parasites are amphipods. *Leucothoe* spp. and other species are commensal rather than parasitic, occurring in the branchial cavity of ascidians or in the canal system of sponges, and the same is true of certain caprellids which are sometimes found on sea stars, and *Hyperia galba*, which sometimes occurs on and in scyphomedusae. Genuine parasites are the Cyamidae or whale lice. They are dorso-ventrally flattened and live, sometimes in large numbers, on the skin of whales and dolphins (Fig. 48).

Insecta. Relatively few insects have invaded the marine habitat (Cheng 1976) and only one is known to be a larval parasite of a marine invertebrate. The caddis fly, *Philanisus plebeius*, oviposits in the starfish, *Pateriella exigua*, which lives in large numbers in intertidal rock pools on the coast of southeastern Australia (Anderson *et al.* 1976, Anderson and Lawson-Kerr 1977). The egg clusters and newly hatched larvae are found in the host's coelom,

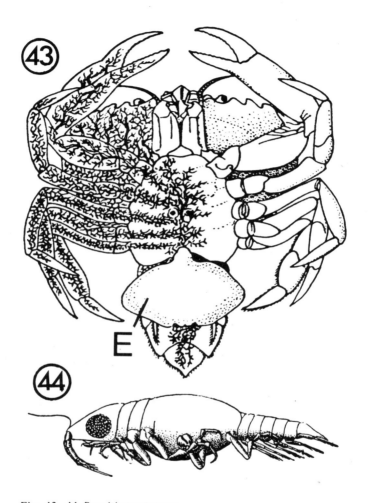

Figs. 43 - 44. Parasitic crustaceans.
43. *Sacculina carcini* (Rhizocephala) in *Carcinus maenas*, ventral view. Parasite drawn only on right side of body (from Kaestner 1959, after Boas). E = external part of parasite.
44. Praniza larva of *Gnathia* (Isopoda) (from Kaestner 1959, after Monod).

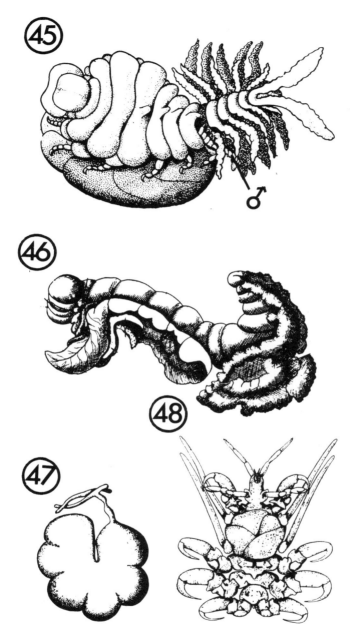

Figs. 45 - 48. Parasitic crustaceans.

45. *Cancricepon elegans* (Isopoda). Ventro-lateral view of ♀ with minute ♂ (from Kaestner 1959, after Giard and Bonnier).
46. *Portunion maenadis* (Isopoda), not fully mature female (from Kaestner 1959, after Giard and Bonnier).
47. Female of *Danalia curvata* (Isopoda) (from Kaestner 1959, after Caullery).
48. Female of *Cyamus erraticus* (Amphipoda), ventral view (from Kaestner 1959, after Iwasa).

where they attack starfish tissue and construct cases from small pieces of ossicle material. The authors suggested that the larvae escape from the host and continue their development in the free environment. However, Leader and Bedford (1979) found that the preferred method of oviposition in New Zealand is in coralline turf. It is not yet clear whether egg laying into starfish is purely accidental.

Insect parasites of marine vertebrates are more common (Murray 1976). Many species of Anoplura (sucking lice) are permanent ectoparasites of seals, whose blood they suck. Those species living on fur seals are always surrounded by an air blanket, whereas those living on hair seals are immersed in water whenever the host is in water. The Mallophaga (biting lice) feed on blood, eggs and moulting nymphs of their own or other species of biting lice, or on feathers and dead skin. Many species infect marine birds including gulls, albatrosses, frigate birds, petrels, etc. An example is the species *Austrogonoides waterstoni* on the little penguin, *Eudyptula minor*, around the southern part of Australia (for further details see Murray 1976).

According to O'Meara (1976), less than 5 per cent of the approximately 2,500 described species of mosquitoes breed in brackish water. Most of these feed on terrestrial vertebrates, but *Uranotaenia latralis* in Malaya feeds primarily on a marine gobiid fish, the mudskipper, which frequently leaves the water.

TARDIGRADA

Only one species, *Tetrakentron synaptae*, a form characterized by a broad dorso-ventrally flattened body, may be a marine parasite. It was found on the tentacles of the sea cucumber, *Leptosynapta inhaerens*, but as its feeding habits are unknown, the nature of the association is not clear (Fig. 49).

PENTASTOMIDA

Adult pentastomids usually parasitize the respiratory tract of mammals, birds and reptiles. Some marine seabirds, such as terns and gulls, are sometimes infected with *Reighardia sternae*, and sea snakes also may harbour pentastomids.

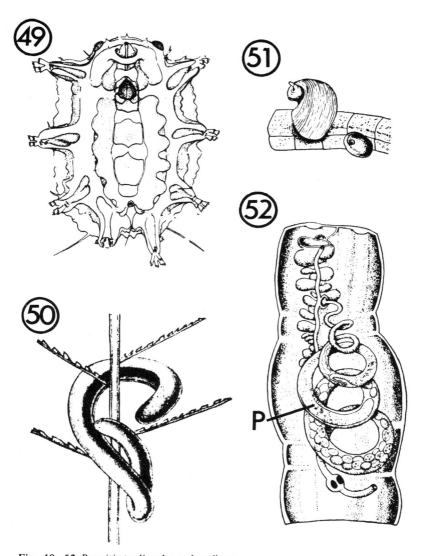

Figs. 49 - 52. Parasitic tardigrades and molluscs.
49. Male of *Tetrakentron synaptae* (Tardigrada), ventral view (from Cheng 1964, after Marcus).
50. *Rhopalomenia aglaopheniae* (Solenogastres) from Kaestner 1955, after Kowalesky and Marion).
51. Female and (smaller) male of *Thyca stellasteris* (Gastropoda) (from Baer 1952, after Koehler and Vanery).
52. *Entoconcha mirabilis* (Gastropoda). Snail attached to the dorsal vessel of host; complementary males are seen in the subterminal swellings (from Baer 1952, after Baer). P = parasite.

MOLLUSCA

The Solenogastres are marine molluscs of a worm-like shape without a shell and distinct head. Several species feed on the polyps and the cenosarc of hydrozoans and octocorallians and are predatory rather than parasitic (Fig. 50). Among the Gastropoda (snails), most parasitic forms belong to the Prosobranchia and infect echinoderms. Some are ectoparasites, for instance *Thyca* on starfish (Fig. 51); others are endoparasitic, for instance the Entoconchidae in holothurians. Certain species of the latter family were found to be connected to the blood system of the host. They have lost all external characteristics of their free-living snail ancestors; only the larval stages are typically snail-like in structure (Fig. 52). *Megadenus* spp. parasitize the respiratory tree of sea cucumbers and their proboscis penetrate into the body cavity or the ring canal of the hosts. Other species of the same genus live in gall-like structures in sea stars and sea urchins. *Stylifer* spp. are partly or deeply embedded in the tissues of sea urchins and starfish, and a considerable number of other genera (*Pelseneeria, Mucronalia, Stylifer, Gasterosiphon, Diacolax, Paedophoropus*) parasitize echinoderms. The opisthobranch family Pyramidellidae (sometimes included in the Prosobranchia) contains species which attach themselves to sessile worms or mussels. They pierce the hosts with a stylet and suck in its juices and tissues (Fretter and Graham 1949). One species of this family, *Odostomia eulimoides*, attacks oysters. The infection often causes death, and the relationship is intermediate between parasitism and predation. A similar relationship exists between snails of the genera *Epitonium* and *Opalia* and coelenterates.

Among the bivalves, species of the superfamily Leptonacea live commensally, often attached to sipunculids, crustaceans, echinoderms and some corals. *Fungicava* lives inside the coral *Fungia*.

No parasites are found among the Scaphopoda and Cephalopoda.

ECHIURIDA

Males of *Bonellia* parasitize the much larger female; this is an example of intraspecific parasitism. All other species are free living.

ECHINODERMATA

A close association with hosts is known from the ophiuroid genera *Asteronyx* and *Gorgonocephalus*, which sometimes feed on the surface layer or the polyps respectively of colonial cnidarians. Young *Gorgonocephalus* spp. appear to feed mainly on the skin of the soft coral, *Gersemia glomerata*. The ophiuroid, *Ophiomaza cacaotica*, is an ectoparasite of tropical crinoids in the Indo-Pacific. Among the sea cucumbers, *Rynkatorpa pawsoni* was found attached to the side of a bathypelagic fish species. But there was no apparent invasion of the host tissue and the association is therefore commensal rather than parasitic (Martin 1969).

CHORDATA

Among marine fishes, lampreys (*Petromyzon* and related genera) and hagfish (*Myxine* and related genera) dig into the host fish and consume their flesh. But as the hosts usually die, particularly if attacked by hagfish, the relationship is predatory rather than parasitic. There is some evidence that the small fish *Carapus*, which lives in association with sea cucumbers, does not only seek shelter in the host when threatened, but actually feeds on its viscera which regenerate, representing a lasting source of food (Fig. 53). The association would, thus, be truly parasitic. The cleaner mimics, *Aspidontus taeniatus*, and other blenniid fish attack fish to bite off pieces of skin or fin and may therefore be considered to be periodic parasites. Finally, some vertebrates are examples of intraspecific parasitism. In some species of the deep sea fish group Ceratioidea, the small males are ectoparasitic on the much larger females. The anterior ends of several males may fuse with the female, sense organs and digestive tract of the males are reduced, but the testes remain well developed (Fig. 54). Snake-eels (Ophichthidae) sometimes are found in the body cavity of other fish, but it is unlikely that they are true parasites. They are probably prey fish which, while trying to escape, bored their way through the wall of the digestive tract and died (Walters 1955). Finally, the skuas and frigate birds chase other birds in flight and feed on their regurgitated food. They are so-called kleptoparasites, "parasites" which live by stealing food.

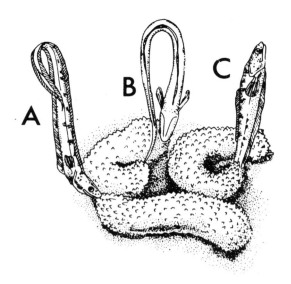

Fig. 53. A fish, *Carapus acus*, that lives inside sea cucumbers. A. Fish searches for anal opening of sea cucumber with head directed forward. B. Introduction of tail into anus of host. C. Pushes itself with tail first into respiratory tree of host (from Osche 1966, after Emery).

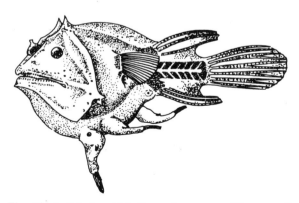

Fig. 54. A fish in which the males are parasitic upon the females. Female of *Edriolychnus schmidti* with three males fused to her ventral side (from Osche 1966, after Günther-Deckert).

CHAPTER 3

The Kinds of Hosts of Marine Parasites

A few examples will be discussed to show that even the smallest animals, and animals belonging to a wide range of phyla and living in a wide range of habitats, may serve as hosts to parasites.

Protozoa, single celled animals, most of which are of microscopic size, are frequently infected by parasites. Doflein and Reichenow (1953) and Dogiel (1965), among others, gave brief accounts of such parasites. Besides symbiotic zoochlorellae and zooxanthellae, as well as fungi and bacteria, which are not of animal nature and therefore not considered here, the list of parasites found in Protozoa includes flagellates, sarcodines, apicomplexans, microsporans, myxozoans, ascetosporans, suctorians and other ciliates, and even nematodes. Protozoan hosts of parasites are flagellates, opalinates, sarcodines, apicomplexans, myxozoans and suctorians and other ciliates. The parasite fauna of marine Protozoa is less known than that of other Protozoa; nevertheless, a number of cases have been studied.

Gregarines of polychaetes, for example, have hyperparasites of uncertain taxonomic position, which are, however, probably fungi (Metchnikovellidae). The ciliate *Phtorophrya insidiosa* is a hyperparasite of another ciliate, *Gymnodinoides corophii*, which itself parasitizes the marine amphipod *Corophium acutum*. Ciliates belonging to the genus *Hypocoma* are ectoparasitic on marine ciliates, the suctorian *Endosphaera engelmanni* infects the ectoparasitic ciliate *Trichodina* and other marine ciliates. Even the minute ciliates belonging to the Tintinnoinea are parasitized by dinoflagellates, and marine dinoflagellates themselves may be hosts to parasitic dinoflagellates. Radiolarians may contain dinoflagellates of the genus *Coccidinium*. Finally, the myxozoan *Ceratomyxa coris*, a species infecting the gall bladder of marine fish, is host to the microsporan *Nosema marionis*.

Bresslau (1928–33) summarized our knowledge of parasites of turbellarians. Larval nematodes and trematodes, a turbellarian, apicomplexans, ascetosporans, ciliates and flagellates have been recorded. Among marine species, trematode larvae were found in various groups of Turbellaria. The turbellarian *Oekiocolax plagiostomorum* parasitizes other marine turbellarians. Apicomplexa and particularly gregarines infect marine polyclads and other turbellarians; one such form is *Lankesteria cyclopori* from the intestine of the polyclad *Cyclophorus maculatus*. The ciliate, *Hoplitophrya uncinata*, lives in the intestine of marine triclads, and a marine *Trichodina* is known from the polyclad *Thysanozoon brocchi*.

One example will show that crustaceans can also be invaded by a great variety of parasites.

Jepps (1937) examined the pelagic copepod *Calanus finmarchicus* in the Clyde Sea area, Firth of Clyde. In spite of its small size (up to 5.5 mm), the host was found to harbour the four species of dinoflagellates *Blastodinium*, *Paradinium*, *Syndinium* and *Ellobiopsis chattoni*, a gregarine, the ectoparasitic ciliate *Chattonella calani*, a parasite of uncertain taxonomic position (*Ichthyosporidium*), two species of larval cestodes and the larva of the nematode *Contracaecum*.

Sprague (1970) briefly discussed protozoan parasites of marine bivalves, some information on parasites of squid can be found in the review by Clark (1966), and Cheng (1967) reviewed the parasites of commercially important marine molluscs. His list shows that the small cockle *Cerastoderma edule* is known to be a host to sporocysts of seven, to rediae of one, and to metacercariae of four species of trematodes; three species of crabs also live in association with it. Lauckner (1971) recovered 13 species of trematodes from two species of *Cerastoderma* in the North and Baltic Seas. According to Cheng (1967), the American oyster *Crassostrea virginica* harbours one species of flagellate, two species of sarcodines, four species of apicomplexans, two species of ciliates, sporocysts of two and metacercariae of two species of trematodes, one larval cestode, two gastropods, two copepods and two crabs. *Mytilus edulis* may be parasitized by the following parasitic species: nine protozoans, ten larval trematodes, two gastropods, four copepods and three crabs.

Altogether, parasites of the following groups have been found in or on marine molluscs of commercial importance: flagellates, sarcodines, apicomplexans, ciliates, microsporans, cestodes, trematodes, nematodes, gastropods, crabs and copepods. To these may be added boring sponges of the genus *Cliona* and cnidarians recovered from the mantle cavity of marine molluscs; these two groups, however, are probably not true parasites. The same refers to nemerteans of the genus *Malacobdella*, several species of which have been reported from the mantle cavity of marine clams.

Harant (1931), in a review of "parasites" of ascidians, listed many species of Protozoa, Nemertini, Turbellaria, Annelida, Gastropoda, Lamellibranchia, Amphipoda, Isopoda, Decapoda and Copepoda. However, many of these are probably not genuine parasites. Parasites of marine fish belong to an even greater variety of groups. Shulman and Shulman-Albova (1953) recorded the following larval and adult parasitic species from 83 individuals of the cod, *Gadus morhua*, in the White Sea: one microsporan, three myxozoans, one ciliate, seven trematodes, three cestodes, four nematodes, two acanthocephalans and two copepods. However, this is only a small proportion of the parasites recorded from *Gadus morhua*. Dollfus (1953), in an extensive critical review, listed 71 parasitic species of this species of fish from various geographical areas, among them ten or eleven nematode species, seven species of acanthocephalans, sixteen species of trematodes, two of monogeneans and nine of cestodes.

Another example from Shulman's and Shulman-Albova's monograph on fish parasites in the White Sea is *Eleginus navaga*. One hundred and forty-three specimens of this species yielded one myxozoan, one ciliate, two monogeneans, eleven trematodes, among them two larvae, two larval and one adult cestode, three larval and two adult nematodes, three acanthocephalans, one leech and one copepod. (Some other examples in Polyanski 1966 and Dillon 1966.)

The examples given are from cold and temperate seas. The diversity of parasites is even greater in warm seas, but no comprehensive surveys considering all groups of parasites have been made. Indications of the great parasite diversity on marine fish of warm water are that 40 specimens of sea bream, *Acanthopagrus australis*, in northern New South Wales were found to be infected with 8 species of Monogenea and 12 species of Copepoda (Roubal 1979). Endoparasites of this fish species have not yet been examined.

In summary, it appears that, although not all groups of animals have been examined with equal thoroughness, parasites (and probably a wide range of them) infect all groups. It may well be that continuing studies will discover parasites for each species in the animal kingdom. The parasite fauna is richest in the host group which contains the largest and most complex species, thereby providing the greatest number of potential "niches" to parasites, that is in the vertebrates (see for instance Dogiel 1964).

The distribution of parasites among the various animal groups shows distinct trends which were emphasized by Dogiel (1964). Deuterostomia, such as Chordata, Chaetognatha, Echinodermata, Hemichordata and Pogonophora, include few genuine parasitic species, whereas groups with a simple body plan and a small size tend to have a greater share of parasitic species.

CHAPTER 4

Parasites of Parasites

A widespread phenomenon in the animal kingdom is hyperparasitism, the infection of parasites by other parasites. It is particularly common among insects of the family Ichneumonidae, although they are parasitoids rather than parasites, that is they kill their hosts after having parasitized them for long periods. Dogiel (1964) provided a brief review. According to him, the species *Apanteles glomeratus*, for instance, has 20 hyperparasitic species, and even fourth degree hyperparasites may be found among the ichneumonids—parasitoids infecting other parasitoids which themselves infect other parasitoids which infect parasitoids of parasitoids in a free-living host.

A number of examples of protozoan hyperparasites of Protozoa were discussed in Chapter 2.

An extensive review of hyperparasites of parasitic helminths was given by Dollfus (1946), who showed that all groups of parasitic helminths examined, trematodes, cestodes, acanthocephalans and nematodes, are hosts to parasites. The hyperparasites found belong to the Microspora, Ascetospora, Mastigophora, Apicomplexa, Opalinata(?), Ciliophora, Trematoda, Cestoda, Nematoda, Protozoa and other parasites of uncertain taxonomic status, as well as algae, bacteria, spirochaetes and many fungi (see also Lee 1971).

Only a few examples of marine hyperparasites can be found in Dollfus' monograph, which may indicate that marine parasites have been much less thoroughly studied than terrestrial and freshwater parasites. Emphasis on non-marine hyperparasites is also indicated by Lee's (1971) review of protozoans and other microorganisms in helminths. Not a single marine form is mentioned.

Marine parasitic crustaceans may be hosts to a number of hyperparasites, some of them also belonging to the Crustacea. The Cryptoniscina, a group of isopods, include such forms. *Danalia curvata*, for instance, lives in the branchial cavity of the marine crab *Inachus* and connects itself by means of its proboscis to the rhizocephalan *Sacculina*; similarly, *Liriopsis pygmaea* parasitizes the rhizocephalan *Peltogaster*. The copepod *Caligus*, very common on marine fishes, may itself carry a monogenean, *Udonella caligorum*, and a ciliate, *Conidophrys* (Dogiel 1964). *Udonella caligorum* has also been found on the copepod *Lepeophtheirus hospitalis* (Schell 1972). Some further examples from Dogiel are: the monogenean *Cyclobothrium* on the marine parasitic isopod *Meinertia* and the copepod *Aspidoecia* on the isopod *Aspidophryxus* which parasitizes marine pelagic mysidaceans.

Among some recent reports of hyperparasitism, all from North America, are those by Sprague (1964) who found the microsporan *Nosema dollfusi* in the trematode *Bucephalus cuculus* from the oyster *Crassostrea virginica* (see also Couch *et al.* 1966), and by Perkins *et al.* (1975), who reported the ascetosporan *Urosporidium spisuli* in anisakid nematodes from the surf clam *Spisula solidissima*. Overstreet (1976) described the myxozoan *Fabespora vermicola* in the trematode *Crassicutis archosargi* from the estuarine fish *Archosargus probatocephalus* (for further records on microsporan and ascetosporan hyperparasites of marine trematodes see Sindermann and Rosenfield 1967, and Canning 1975).

Altogether, it appears that parasites of many (and probably all) groups of animals may be hosts to hyperparasites, but relatively few hyperparasites in the marine environment are known. This may be due in part to the relatively small number of detailed studies of marine parasites, but a more important reason is perhaps that there are more hyperparasites, and hyperparasites of higher degrees, in groups like the ichneumonids than in marine animals.

CHAPTER 5

General Adaptations of Parasitic Animals

All organisms are adapted to their "niche" (see Chapter 7), and because many parasites live under extreme conditions in or on their hosts, some of their adaptations are extreme. For example, the louse *Antarctophthirus ogmorhini* lives on the Weddel seal, *Leptonychotes weddelli*, in Antarctica. For survival, it must be able to withstand the very low temperatures and the high pressures at several hundred metres depth to which its host dives. It has been shown that the louse survives supercooling to $-20\,°C$ for 36 hours, and diving to 600 m for 45 minutes (Murray 1976). Besides such special adaptations, many parasites have certain characteristics in common which can be interpreted as adaptations to the parasitic way of life in general. Such adaptations in the various animal groups were discussed by Baer (1952) and Dogiel (1964), mainly from a morphological and ecological, and by Rogers (1962) from a physiological point of view.

Size of Parasites

Parasites can avoid excessive damage to and death of their hosts only by remaining relatively small. There are few parasites which approach their hosts in size. Noble and Noble (1976) estimated the length of a didymozoid trematode in a sunfish, *Mola mola*, to be 12 m. Figure 55 shows a large nematode in a whale. In spite of their length, both parasites are much smaller than their host if the comparison is based on volume or weight. Some ectoparasitic isopods of small fish may reach one quarter to one half the volume of their hosts, and lampreys may be as large or even larger in volume and weight than their hosts. The latter, however, are often predators rather than parasites, often killing their prey. Protozoan parasites of Protozoa are commonly relatively large.

In spite of the small size of parasites compared with the size of their hosts, they are often much larger than their free-living relatives.

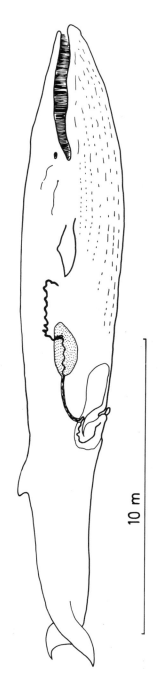

Fig. 55. One of the largest parasites. Reconstruction of female of the nematode, *Crassicauda crassicauda*, in the urinary ducts and blood vessels of a fin whale (reconstruction and drawing by Gibson, British Mus. Nat. Hist.).

Many nematodes and cestodes may serve as examples. One reason may be the necessity to produce large numbers of offspring.

Increase in reproductive capacity

Many authors have drawn attention to the large number of offspring produced by many species of parasites. Useful in this context is the distinction between r- and K-selection. Animals which are r-selected, invest much of their energy in producing many offspring (population size is determined by the intrinsic rate of population growth r), and little energy is spent on maintaining the individual (individuals have a relatively simple structure). In K-selected animals, the reverse is true. Much energy is spent on maintaining the individual (with high complexity) and less on producing offspring (population size is determined by the carrying capacity of the environment K). There is a whole spectrum of possibilities between extreme r-strategists and extreme K-strategists, but parasites, in general, are more r-selected than their free-living relatives. For instance, Jennings and Calow (1975) have shown that free-living flatworms have a high energy content but produce few eggs; they are the most extreme K-strategists in that group. Ectocommensal flatworms have a slightly lesser energy content and usually produce rather more eggs, ectoparasitic flatworms are even poorer in energy and produce even more eggs, and endoparasitic flatworms (trematodes and cestodes) are extreme r-strategists, producing very large numbers of eggs and containing little energy. In addition, the number of offspring is increased by parthenogenetic or asexual multiplication of the larvae. Table 2 gives some estimates on the reproductive potential of the various groups of Platyhelminthes.

Some parasites produce truly enormous numbers of offspring. Thus, one female of the parasitic neorhabdocoel turbellarian,

Table 2. Approximate estimates of fecundity of free-living and parasitic Platyhelminthes. Data on eggs from Jennings and Calow 1975, larvae author's own estimates.

	Number of eggs during life of a worm	Multiplication of larval stages
Turbellaria	10	$\times 1$
Monogenea	1,000	$\times 1$
Digenea	10 mill.	$\times \geqslant 1,000$
Cestodes	10 mill.	$\times (1-1,000)$

Kronborgia caridicola, lays more than one million eggs during its life (Kanneworff and Christensen 1966). The protozoan *Nematopsis ostrearum*, which uses molluscs and crabs as hosts, may release as many as 8 million cells of the stage infective to molluscs, between two moults of its crab host (Prytherch 1940).

Nevertheless, not all parasites produce such large numbers of offspring, and even within one group there is a great variability. This is shown by the Monogenea. Freshwater monogeneans, which are known best, produce from less than 100 to thousands of eggs during their life (data from Gröben 1940, Bychowsky 1957, Kohlmann 1961), and it is probable that marine monogeneans have a similar range in egg production.

However, values for egg production and multiplication of larval stages alone do not permit an estimate of fecundity. As pointed out by Kennedy (1975, 1976b) and others, generation time also has to be considered, being the time from formation of the egg to development of a mature female. Fecundity of a species with a small number of eggs and a short generation time can equal that of a species with a large number of eggs but a long generation time. Species of the monogenean *Gyrodactylus*, for instance, produce relatively few eggs, only one fertilized egg cell being present at any time in the uterus, but the generation time is extremely short; the egg develops in the uterus to a daughter animal which already contains a developing embryo of the next generation. Furthermore, the life span, or rather the duration of reproductive activity, is important. A long lived animal with slow reproduction may have greater fecundity than a short lived one with fast reproduction. All factors contributing to fecundity vary between individuals of a population as well as with time and are difficult to determine. Hence, as stated by Kennedy: "the actual, as opposed to the potential number of parasite progeny produced by a *population* has seldom been determined under natural conditions", and such data are not available for any species of marine parasites.

Two opposing trends: reduction and increase in complexity

In the foregoing paragraph, I have shown that many parasites produce more offspring than their free-living relatives and thereby overcome the hazards implicit in life cycles which depend on finding a host. Corresponding to this high reproductive capacity is thought to be a reduction in structural complexity, most clearly seen in the reduction of sense organs and nervous systems. Dogiel (1964) for instance, wrote in his authoritative treatise on general parasitology (p. 143):

The nervous system and the sense organs require no long discussion. All parasites, especially the endoparasites, have a more or less simplified nervous system. This is best reflected in the loss of the sense organs, which can be present in the free-living developmental stages and disappear with the transition to the parasitic way of life.

Structural degeneration of parasites, as compared with their free-living relatives, does indeed often occur. A good illustration of this is the various species of Rhizocephala (for instance *Sacculina*), extremely modified barnacles that live on and in marine and freshwater crustaceans. The larval stages have all the characteristics of larval barnacles, whereas the adults resemble tumour-like growths in the tissue of the host, without much external organization, and without well developed nerves and sense organs, although the presence of some sense receptors is indicated by reaction of the external parts to stimuli (Fig. 43). Another example is the isopod *Danalia curvata*, which is hyperparasitic on *Sacculina*. *Danalia* is first a male but becomes a female at a later stage of development, the latter consisting mainly of a sac-like structure containing the gonads, attached to its host by a long proboscis, and with a complete reduction of mouth parts, limbs and their associated nervous and sensory structures (Fig. 47). In the females of the ascothoracid crustaceans *Myriocladus* and *Dendrogaster* in the coelom of sea stars, the shape is completely determined by lobe-like outgrowths of the mantle, with a strong reduction of most organs (Fig. 42).

A careful analysis shows, however, that in ecto- and endoparasites there is frequently not only no reduction of nervous and sensory structures, but on the contrary they exhibit an increased complexity. This is true even in some trematodes, the group chosen as an example of simplified nervous systems by Dogiel. Marine species have not been studied in detail; therefore freshwater species with close relatives in the sea will be discussed. Rohde (1966, 1968a,b,c,d, 1970, 1971, 1972a,c) analysed the nervous system and sense receptors of an aspidogastrid trematode, *Multicotyle purvisi*, and of a digenean, *Diaschistorchis multitesticularis*. Both species are endoparasites of turtles, and both were shown to have nervous systems of extraordinary complexity (Fig. 56). In *Diaschistorchis* the high degree of complexity can be considered to have been secondarily acquired because it is "rare among Digenea and appears to be connected with the presence of a head collar which is clearly a derived character". In adult *Multicotyle*, six to eight distinctly different types of sense receptors were found, some of them located only on certain parts of the body, and parts of the posterior ventral nerve cord were shown to be surrounded by a nerve sheath, a structure not known from any

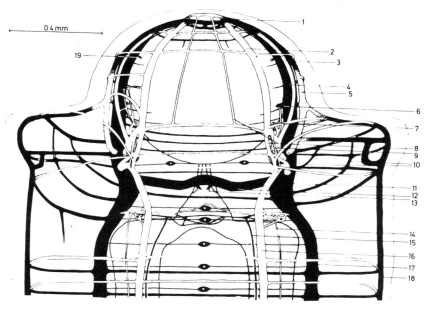

Fig. 56. Anterior part of nervous system of *Diaschistorchis multitesticularis*, dorsal view (after Rohde 1968a). Numbers refer to various nerves.

other lower invertebrate. Large numbers of sense receptors were shown to occur in a marine species related to *Multicotyle* (Rohde 1972b), and a nervous system of great complexity was also found in a monogenean, *Polystomoides malayi*, an endoparasite in the urinary bladder of turtles (Rohde 1968b, 1975a). Electron microscope studies revealed that the parasite not only has simple ciliated sense receptors, but also possesses compound receptors consisting of bundles of axons with ciliated endings. In a review of receptors in Monogenea based on her own work and that of others, Lyons (1973) described several types of sensory structures, some shown to occur in very large numbers (Fig. 57). Parasitic nematodes, in spite of their structural uniformity, also have a wide range of receptors, as shown in the recent reviews by McLaren (1976) and Wright (1980).

These examples show clearly that two ways are open to a parasite. The first way is reliance on the host. The parasite allows the host to take over some of its functions and pays for it by the loss or reduction

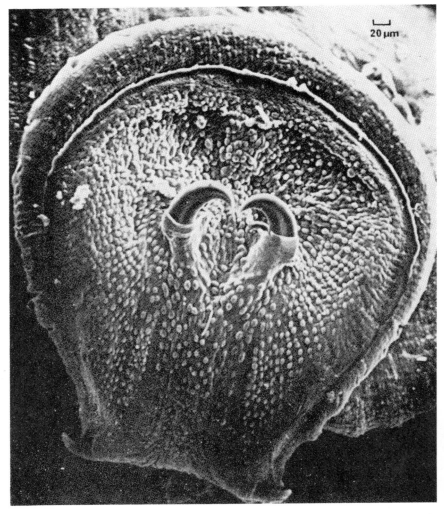

Fig. 57. Scanning electron micrograph of opisthaptor of *Entobdella soleae*. Note the numerous sensory papillae (after Lyons 1973).

of certain organ systems. The second way is to compensate for the reduced external stimuli by an even greater development of nerves and sense receptors. This would enable the parasite to find favourable niches within the host, for instance with regard to acidity, osmotic pressure, oxygen pressure, a certain tissue texture, etc. Unfortunately, little is known about how parasites find their niches in or on the hosts and practically nothing is known about the function of sense receptors and nervous systems in parasitic lower animals.

Dispersal

Dispersal is important for the survival of any species because a population restricted to one small area risks becoming extinct, if conditions become unfavourable, and because dispersal reduces inbreeding and the loss of evolutionary adaptability. It also prevents overinfection of hosts (Kennedy 1975). Mechanisms ensuring dispersal of parasites differ in some ways from those of free-living organisms. It is convenient to follow Kennedy (1976b) and distinguish the following components of dispersal in parasites: (1) dispersal away from an individual host, (2) dissemination in space and range extension, usually over wide distances and (3) dispersal in time.

The first component, dispersal away from a host, is often achieved by actively swimming stages, for example cercariae (Fig. 58) and miracidia of trematodes, or coracidia of certain cestodes, which are also infective stages. In other words, it is difficult to judge how much of the swimming activities actually have the function of infecting another host, and how much of it is related to dispersal. The same applies when larvae with little active locomotion of their own are dispersed over short distances by their host. For example, several hundred cercariae of *Bucephaloides gracilescens*, which are not active swimmers, are forcibly discharged at the same time by their mollusc host, *Abra abra* (Matthews 1974). This assists in their dispersal but it probably also enhances the chances of coming into contact with their next hosts, marine fishes. For dissemination in space, a parasite usually depends on its host, because most hosts have a greater motility than their parasites. Dispersal in time may be brought about by a long life span of the parasite in the host and production of eggs or offspring over a long time, or by resting stages such as eggs or cysts either within the host or in the free environment. Frequently, all three components are combined. An example may demonstrate this. The trematode, *Austrobilharzia terrigalensis*, lives in the blood systems of

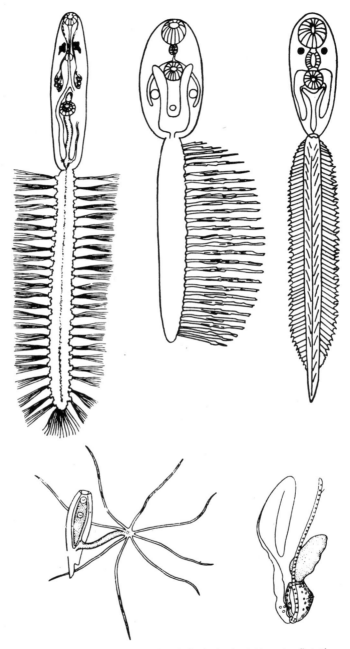

Fig. 58. Marine cercariae. Morphological adaptations to flotation (from Ginetzinskaya and Dobrovolski 1978, after Cable, Tschubrik, and Chabaud and Bigue).

silver gulls and reef herons on the Great Barrier Reef (Rohde 1977d). As indicated by the long life span of other schistosomes, it probably lives for several years, continuously producing eggs. Infected birds spread the eggs in space because of their vagility, and they spread them in time because of the prolonged production of eggs. The miracidia in the eggs infect snails of the species *Planaxis sulcatus*. Larval development in the snails continues at least for several months and perhaps years and leads to the continual release of actively swimming cercariae from the snails, which further contributes to dispersal in time. The cercariae disperse around the snail host, but only over short distances, because their life span is short and their swimming activities are weak. However, tidal currents may contribute to wider dispersal of this stage around a particular island, but they are probably not sufficient to carry it to other islands.

Sometimes, dispersal depends entirely on the host. Thus, the aspidogastrid *Lobatostoma manteri* lives as an adult in the fast swimming fish, *Trachinotus blochi*, on the Great Barrier Reef. It produces eggs over long periods and dispersal in time and space is thus ensured. The eggs are eaten by snails, *Cerithium moniliferum*, which are sluggish and move around very little, and infection of the fish host occurs by ingestion of the snails. Altogether, dispersal is entirely due to the movements of the fish (Rohde 1972d, 1973, 1975b).

Dispersal by inactive resting stages occurs, for instance, in many trematodes. The metacercariae are encysted in marine pelagic invertebrates like Chaetognatha, or vertebrates such as fish. Final hosts become infected by eating the hosts carrying the larvae, at different times and in different places.

Sometimes dispersal is helped by man. For example, the copepod *Mytilicola intestinalis*, which infects various bivalves, has some free-living stages, but they are only of a few days' duration. Dispersal by them over wide distances is probably not common, and introduction into new areas is apparently usually brought about by live mussels on the bottom of ships, as indicated by the fact that isolated infections repeatedly have been found near ship breaking and ship repairing yards (several authors; see Hepper 1955).

Mechanisms of infection

Survival of parasites depends on their success in infecting the right host(s). Corresponding to the wide range of parasites is a wide range of infection "strategies". Infection may be by contact transfer, by ingestion of transport (paratenic) hosts which carry infective larvae, by eggs, spores or cysts, or by free-living larvae. Examples of contact transfer of parasites from one individual host to another are monogeneans of the family Gyrodactylidae, which infect freshwater and marine fishes (Bychowsky 1957, Khalil 1970). Transport hosts, many species of fish and invertebrates, are used by larval anisakid nematodes. Final hosts, which are often marine mammals, become infected by eating the transport hosts. Bychowsky and Nagibina (1967) presented convincing circumstantial evidence, but still without experimental proof, that fast-swimming predatory pelagic fish of the family Cybiidae become infected with the monogeneans *Pricea* and *Gotocotyla* by eating smaller fish of several families which serve as "intermediate" hosts. Large numbers of small larvae which do not develop beyond a certain stage are found throughout the year on the gills of these fish. They apparently migrate actively to the gills of the "final" host when eaten. Direct infection of "final" hosts with free-swimming larvae is also thought to occur, but much more rarely.

Eggs, spores or cysts often can resist adverse environmental conditions for long periods, and they usually have a low rate of metabolism and do not develop outside the host. Hatching of eggs or cysts which enter the host through the mouth may occur in several ways (Rogers 1963, and review by Lackie 1975). Firstly, the host may stimulate the secretion of digestive enzymes by the egg or larva, which leads to the dissolution of the shell or cyst wall. Secondly, the host may induce an increased activity of the larva which leads to rupture of the shell or cyst wall. Thirdly, there may be direct digestion of the shell or cyst wall by digestive enzymes of the host. All of these processes, or only one or two, may be important in a particular species. Some host factors involved are pH, temperature, certain enzymes and bile, but all examples examined so far (except for helminths of some sea birds) are non-marine.

The environment provided by the host itself may also be involved in hatching of infective stages which do not enter the host orally, and often such factors are superimposed on an endogenous hatching rhythm. For instance, hatching in several species of marine Monogenea was investigated by Kearn (1973, 1974a,b, 1975a,b), Kearn and MacDonald (1976) and MacDonald (1974). Urea in the skin mucus of *Solea solea* stimulates production of a proteolytic hatching fluid by the larva of *Entobdella soleae* and thus hatching. Washings from other fishes may also induce hatching, that is the hatching factor is non-specific. Hatching induced by urea is superimposed on an endogenous hatching rhythm, which leads to hatching during the first hours of a period of illumination. Urea from the skin of rays, but not from *Solea*, was shown to release hatching in the monogenean of rays, *Acanthocotyle lobianchi*, but mucus did not lead to hatching in *Entobdella hippoglossi* on the skin of *Hippoglossus hippoglossus*, larvae of which hatch mostly at the end of the first two hours of a period of darkness. *Acanthocotyle lobianchi*, in contrast, cannot be stimulated to hatch by alternating periods of light and darkness, hatching occurring almost immediately when skin mucus from the hosts *Raja* spp. is added, and *Dictyocotyle coeliaca*, a monogenean from the body cavity of *Raja naevus*, has neither an endogenous hatching rhythm nor can hatching be released by host skin mucus or fluid from the host's body cavity. According to Kearn (1975a), absence of an endogenous hatching rhythm may point to the lack of a well defined rhythmical daily activity pattern of the host.

MacDonald (1975) found that the eggs of three species of the monogenean *Diclidophora* have different hatching rhythms adapted to different behavioural patterns of their hosts. *Diclodophora merlangi*, *D. luscae* and *D. denticulata* infect marine whiting (*Merlangius merlangus*), pouting (*Trisopterus luscus*) and coalfish (*Pollachius virens*) respectively. Eggs of the first species from Arbroath, Scotland, hatched mainly in the first 4–6 hours of illumination, but eggs of the same species from Plymouth, England, hatched mainly in the 2 hour period before illumination. Eggs of *D. luscae* hatched over dusk, and eggs of *D. denticulata* hatched after the light was switched off. These differences correspond to differences in the behavioural patterns of the host fish. Neither mechanical disturbance nor the proximity of host tissue caused hatching in the first two species.

If there are free infective stages, infection may be purely passive. For instance, free-swimming ciliates are passively imbibed by their host lamellibranchs (Fenchel 1965). Coracidia larvae of tapeworms, apparently are not attracted by their hosts, and infection appears to be passive (Ulmer 1970). However, often the infective stages have behavioural mechanisms which bring them into contact with the host. For instance, miracidia of certain trematode species are chemically attracted by their mollusc host (Ulmer 1970) and response to light is likely to contribute to infection in the copepod *Pachypgus gibber*, a parasite of the ascidian, *Ciona intestinalis*. Young stages are attracted by light which could result in their moving into the water column where they would be carried to potential hosts. But there is a change from positive to negative phototaxis at a stage of development when penetration into the host, where light is reduced, probably occurs. Furthermore, infective larvae have more winding and slower swimming movements than earlier larvae, which would increase their chances of contacting a host (Hipeau-Jacquotte 1977). The fish *Carapus bermudensis*, which withdraws into the sea cucumber, *Actinopyga agassizi*, when in danger, apparently uses chemical stimuli to locate its host. At least in the laboratory, chemical stimuli of host origin are effective in leading the parasite to the fish (van Meter and Ache 1974). Sastry and Menzel (1962) demonstrated that the crab *Pinnotheres maculatus* is actively attracted by its mollusc hosts, although the attractant could not be identified, and according to Humes (1942), larvae of the nemertean *Carcinonemertes carcinophila* have a positive phototaxis which leads them to the surface and into contact with their crab host.

Occasionally, an infective larva mimics a free-living food animal of its host and thus enhances its chances of being eaten. Pearson (1968) suggested this for the cercaria of *Paucivitellosus fragilis* at Heron Island, Great Barrier Reef. In shape, colour and behaviour the cercaria resembles free-living worms which are probably part of the diet of its final host, a blenniid fish. Fish become infected by eating the cercaria.

Once the host is found, infective stages may need some mechanism which ensures adhesion to the host. Kearn (1974c) described various types of gland cells in the oncomiracidia of three species of the marine monogenean *Entobdella*, some of which may produce an adhesive secretion. The larvae attach themselves first by the head region, then the opisthaptor unfolds and becomes attached by means of its hooks while the head becomes detached (Kearn 1967, 1970b). The cercaria of *Transversotrema patialensis* has specialized regions of the tail, the "arm processes", which apparently are adapted to recognition of the right host and its rapid attachment to it. Adhesive pads on the arms are brought into contact with the host's skin and an elaborate behaviour pattern of the cercaria ensures firm attachment (Whitfield *et al.* 1975).

Parasites may induce remarkable behavioural changes in hosts which facilitate infection of the next host. Such a behavioural change was described by Sparks and Chew (1966) in the clam, *Venerupis staminea*, induced by larvae of the cestode *Echeneibothrium*. Healthy clams typically spend their entire post-larval life buried, whereas infected ones were found exposed on the surface. It is possible that this facilitates completion of the parasite's life cycle by making the host more easily found and eaten by skates or rays. Several examples of parasite induced behavioural changes in intermediate hosts which enhance the chances of infecting the final hosts were discussed by Holmes (1976), but all examples were non-marine.

Sometimes, chances of infecting a host are enhanced by synchronization of parasite reproduction and host behaviour. Thus, the monogeneans *Gastrocotyle trachuri* and *Pseudaxine trachuri* infect their host, *Trachurus trachurus*, near the sea bottom. In summer, the fish disappear from the sea bottom to feed on plankton. The parasites have adapted themselves to this by ceasing to produce larvae in anticipation of the summer migration (Llewellyn 1962). Kennedy (1975) described another example. Shad, an anadromous fish, is dispersed in the sea for most of the year, and there is no contact between adult and young fish. In May, however, such contact is established when fish mass off estuaries. The monogenean of shad, *Mazocraes alosae*, produces rapidly hatching eggs only in May, when infection of young fish is possible.

Aggregation

The distribution of organisms in space may be even, random, or aggregated. The last is also referred to as clustered, overdispersed, or contagious. In evenly distributed organisms, the distances between neighbouring individuals in a population are more or less identical, in randomly distributed organisms they are random, and in aggregated distributions organisms form clusters within their area of distribution. Most parasites have aggregated distributional patterns within their host populations, that is many host individuals are empty whereas a few are more heavily infected than would be expected if the distribution were random (Crofton 1971).

Examples can be found in all groups of parasites. Shotter (1976) found that most helminth species in the whiting, *Merlangius merlangus*, had an aggregated distribution, Pearre (1976) demonstrated overdispersion for trematode larvae in Chaetognatha and Bortone et al. (1978) showed it for the two copepod species *Bomolochus concinnus* and *Ergasilus manicatus* on the two marine fish species *Menidia beryllina* and *M. peninsulae*. But there are exceptions. Thus, according to Kennedy (1979), the cestode *Eubothrium parvum* in capelin, *Mallotus villosus*, is aggregated in its host in a certain locality, but not in another.

Aggregated distributions are so characteristic of parasites that Crofton (1971) even used it for a definition of parasitism. However, it must be kept in mind that such distributions are also common among free-living animals and plants.

Conditions which lead to aggregation in parasites are, according to Crofton (1971):

1. A series of exposures, but each exposure has a different chance of infection;
2. The infective stages are not randomly distributed;
3. An infection increases the chances of a further infection;
4. An infection decreases the chances of a further infection;
5. Variations in host individuals make the chances of infection unequal;
6. The chances of infection of individual hosts change with time.

Most of these conditions usually are jointly responsible for aggregation. Thus, it will usually be the case that numbers of cercariae to which fish are exposed vary at different exposures (condition 1), and rarely or never will infective stages be randomly distributed (2). A previous infection sometimes leads to a decrease in the host's resistance to a subsequent infection (3), but on the other hand an infection may also immunize a host (4). Rarely will individuals of a host population be so uniform that chances of infection are equal; it is more probable that they will show more or less marked differences in resistance to an infection, or in feeding habits which lead to different infections, etc. (5). The susceptibility of a host to a parasite may increase or decrease with age, there may be a seasonal change in the numbers of infective stages, or changes with age or season may be due to a changed diet or behaviour, etc. (6). To these six points another may be added: a single parasitic individual, once on or in a host, may multiply and lead to heavy infection of the same host individual (for example, monogeneans of the families Gyrodactylidae and Ancyrocephalinae, and see below).

Other authors have suggested further conditions which may lead to aggregated distributions. According to Pearre (1976) Mediterranean Chaetognatha prefer food which contains trematode larvae, and such selection is partly responsible for aggregation of trematode larvae in them.

May (1977) suggested as a possible biological function of aggregation the stabilization of the host–parasite association. According to Kennedy (1975), because of overdispersion only few heavily infected hosts of a population die. Death of these hosts may ensure completion of the life cycle of a parasite which depends on being eaten by a second host, but the host population as a whole is not greatly affected. Kennedy (1976a) drew attention to the increased chances of mating in aggregated populations. If parasites were randomly dispersed, the majority of hosts would contain only a single parasite individual and the probability of mating would be very low. However, suggestions concerning the biological function of over-dispersion are purely speculative and no evidence supporting them is available.

Hermaphroditism, parthenogenesis and asexual reproduction

A major difficulty for parasites is to get to a host. Contact is usually sporadic and only a few parasite individuals succeed in infecting a host. It would be of obvious advantage if a single individual could build up a new population once in or on the host. There are several mechanisms which achieve just this: hermaphroditism, parthenogenesis and asexual reproduction (and among non-marine arthropods, dispersal after mating, Price 1980). Hermaphroditism is occurrence of both sexes in one animal, parthenogenesis is development of offspring from unfertilized egg cells, and asexual reproduction is development of offspring from cells other than egg cells and without fertilization. The discussion by Suomalainen (1962) shows that transport of a single parthenogenetic female in any stage of development is sufficient for colonizing a new area, or in the case of parasites, a new host individual. And the same is true for asexual reproduction. Hermaphrodites can successfully build up a population from a single individual only if they are capable of self-fertilization. But even if cross-fertilization, mutual fertilization of two individuals, is obligatory in hermaphrodites, the success of contact between individuals is increased by a factor of at least two compared with bisexual forms, because the partner will *always* have fertilizing capabilities (Tomlinson 1966). In such cases colonization of a new host individual is much easier than in bisexual populations.

There are indeed many examples of hermaphroditism in parasites, including almost all digenetic trematodes, all Monogenea, Aspidogastrea and cestodes. Parthenogenesis (and possibly asexual reproduction) leads to multiplication of larval stages of trematodes inside the mollusc intermediate host, and asexual reproduction by binary or multiple fission is common among parasitic Protozoa.

However, it must be pointed out that evolution of hermaphroditism, parthenogenesis and asexual reproduction is favoured not only in parasites but in other organisms which occur at low densities and have difficulty contacting each other because of low mobility, or for other reasons (Tomlinson 1966). These mechanisms are also advantageous in unstable environments that are frequently depopulated (Stalker 1956).

Ghiselin (1969) suggested that hermaphroditism does not only evolve when it is hard to find a mate, but also when there are small, genetically isolated populations, and when one sex benefits from being larger than the other. In small populations, sequential hermaphroditism (one sex maturing before the other) may reduce the probability of inbreeding among siblings, because if all members of a brood grow at the same rate and pass simultaneously through male and female stages, they are not able to crossbreed. Simultaneous hermaphroditism (simultaneous maturation of both sexes), on the other hand, maximizes the effective population size by making the numbers of males and females nearly equal at any particular time. The sex ratio deviates considerably from equality in many animal groups and particularly in small populations.

Females may benefit from being larger than males as they produce eggs which are usually larger than sperm cells; the larger body size of the female would thus facilitate egg production. Cestodes exhibit protrandry, that is, early maturation of the male reproductive system in the more anterior small segments, and late maturation of the female system in the posterior large ones; the female system may benefit from being in the larger segments.

Many populations of parasites occur in small populations isolated from those in other hosts (see Chapter 7). Ghiselin (1969) believed that low population density is responsible for hermaphroditism in parasites. However, it seems that small size and isolation of populations as well as size benefits to one sex also play a role, although it is difficult to decide how important the various components are.

Price (1980) claimed that parthenogenesis, besides enabling a single parasite individual to colonize a new host, has another advantage: it permits maturation at an early larval stage (progenesis), because adult characters required for mating are unnecessary. Rapid maturation may lead to an increased rate of population growth and a small size of reproducing animals, and both are advantageous to most parasites. That this mechanism may be important is indicated by trematodes. Some metacercariae have become progenetic without reduction of their reproductive systems. However, the simple sporocysts and rediae can be interpreted as being progenetic, parthenogenetic stages. These are universal among trematodes.

Hermaphroditism, parthenogenesis and asexual reproduction do not only have advantages, but they are also thought to have one major disadvantage; they are thought to lead to reduced variability and adaptability and altogether to a slowing down of evolution (Suomalainen 1962). The disadvantage due to parthenogenesis is, however, not as great as usually assumed. Price (1980), among others, pointed out that parthenogenesis does not reduce mutation rates and that parasites produce many offspring and, hence, probably have sufficient mutants. Furthermore, according to Price (1980), a reduced variability may be of advantage for several reasons. Firstly, the genetic make-up of hosts changes relatively slowly because of longer generation times compared with those of their parasites, and hence a slowing down of genetic changes of the parasites is advantageous to ensure coevolution with the host. Secondly, parasite populations fluctuate greatly and a reduced variability guarantees that long term adaptations to the hosts are maintained during "crashes" and the subsequent build-up of populations. Furthermore, the parasites' environment is often more uniform than that of free-living forms and does not demand as great a variability as the environment of free-living organisms.

Altogether, natural selection, depending on the specific conditions, will sometimes favour bisexuality, but sometimes one or all of the other modes of reproduction.

"Goodness" of parasitism

Phoresis, commensalism, mutualism and symbiosis are closely related to parasitism (see Introduction). It is often impossible to draw a clear line between these phenomena and parasitism, in many cases because of insufficient knowledge, but in others because no genuine and clearcut border exists. Best studied are parasites of man, and for several species it has been shown that they live as commensals, feeding on bacteria and detritus in the gut under certain conditions, whereas they become parasites feeding on host cells under other conditions (for example *Entamoeba histolytica*, a protozoan causing amoebic dysentery). Furthermore, certain species may be harmful to one host species, but harmless to another. For this reason, some authors use the term symbiosis for all close associations between organisms, without trying to put an association into any particular category (Read 1958, Cheng 1967). Other authors (Smyth 1962, Lincicome 1963, 1971) prefer parasitism as the general term because symbiosis has been used commonly for organisms living in a close and obligatory association which is of benefit to both partners. Smyth (1962) stated that mutualism and symbiosis are merely special cases of parasitism, in which some metabolic byproducts of parasites are used by the host, and Lincicome (1963) viewed parasitism as a "fundamental expression of a chemical (or molecular) relationship between two organic beings", which may vary in its degree of dependence, the harmoniousness of the relationship, the degree of antagonism, etc.

Lincicome (1971), in particular, drew attention to the fact that parasites often contribute an "element of goodness" to the relationship. He made detailed, carefully controlled experimental

studies of three parasitic species, the protozoans, *Trypanosoma lewisi* in rats and *T. duttoni* in mice and the nematode *Trichinella spiralis* in rats. For all three host–parasite systems, he found that the host benefited in some ways from its parasite, as indicated by the following:
1. All parasitized animals grew at a faster rate; in other words, they were bigger and apparently healthier;
2. Animals fed with a diet deficient in certain factors grew better when they contained parasites;
3. Animals infected with trypanosomes (and under certain conditions with *Trichinella*) ate more food;
4. Activities of certain enzymes in the kidney and liver were reduced by diet deficiency, but elevated when animals were infected, indicating that the parasites restored the missing factor or supplied a similar substance;
5. Panthothenate (vitamin B_2) and pyridoxine levels in the liver of rats infected with trypanosomes or *Trichinella* were raised; both substances are important to the host;
6. Thiamine (vitamin B_1) deficient rats infected with *Trypanosoma lewisi* and pantothenate deficient mice infected with *T. duttoni* lived significantly longer than uninfected animals.

The parasite load was kept small in an attempt to simulate natural conditions.

In addition to these carefully conducted quantitative experiments, uncontrolled daily observations in the laboratory left the impression that infected rats and mice were more active physically, and were more responsive to the human presence.

To evaluate the contribution of the host to the relationship, parasites were studied in the normal hosts in which they grow, and in abnormal hosts where they ordinarily do not. Lincicome could demonstrate that *Trypanosoma lewisi* is dependent upon a certain factor contained in the blood serum of its normal host, the rat. Only if this factor was introduced into mice did the parasites grow in this abnormal host. Even after 300 passages through mice over three years the parasite was still dependent on the rat factor. It could also be shown that the factor did not change the defence reactions of the host to the infection but was required for nutrition of the parasite.

These experiments show that even in supposedly clearcut cases of genuine parasitism, the relationship sometimes (and perhaps as a rule?) may be what is usually called mutualistic, that is of some benefit to both partners.

Even if a parasite does not benefit an individual host, it seems possible that it may be beneficial to the host species by eliminating the weaker host individuals from a population. One might suspect that such effects are widespread, but little or no direct evidence is available.

Among marine parasites, nothing is known of potential mutual beneficial effects of hosts and animal parasites, although Berland (1980) suggested that ascaridoid nematodes in the stomach of fishes, marine birds, seals and whales may mechanically loosen and break up large food particles, permitting the digestive fluids to seep into the core quickly. Such action could be important in digestion, because many hosts ingest their food whole or in large chunks.

CHAPTER 6

Host–Parasite Interactions

Some parasites and hosts are engaged in a continuous battle to take advantage of each other. Parasites attempt to gain entrance into a host and utilize its resources, and their hosts—in turn—attempt to prevent infection, get rid of the invaders or at least limit the damage done by them. Mechanisms employed by the hosts for this purpose are behavioural, or they are immunity and tissue reactions.

Behavioural reactions of hosts to parasites

Animals employ various behavioural methods to free themselves of parasites. Activities of birds which are thought to result in the removal of ectoparasites are preening, bathing in water or dust, and passive as well as active "anting" in which birds bathe in or cover themselves with ants. Mammals rub their bodies against hard objects or are "cleaned" by birds. The latter is an example of a cleaning "symbiosis", a relationship in which both partners derive a benefit; the cleaner acquires food and the host has dirt particles and parasites removed from its body surface.

Aquatic and particularly marine animals use similar methods. The marine fish *Caranx chrysos* and *Elagatis bipinnulatus*, for instance, were seen to scrape their skin against sharks (Eibl-Eibesfeldt 1955), and Fiedler (1964) saw wrasses (Labridae) of the genus *Crenilabrus* use jerky movements of the fins, shaking of the head and tail, and rubbing of the body against edges of a rock, apparently to get rid of ectoparasites; repeated strong spitting may have the function of removing parasites from the gills and mouth cavity. Fish may jump out of the water, dive rapidly or swim erratically apparently to remove ectoparasites (Kollatsch 1959, Hotta 1962), and it seems also possible that secretion of mucus envelopes by many species of Labridae (wrasses) and Scaridae (parrotfish) at night, in some cases only by juveniles (Winn 1955, Winn and Bardach 1960) has the

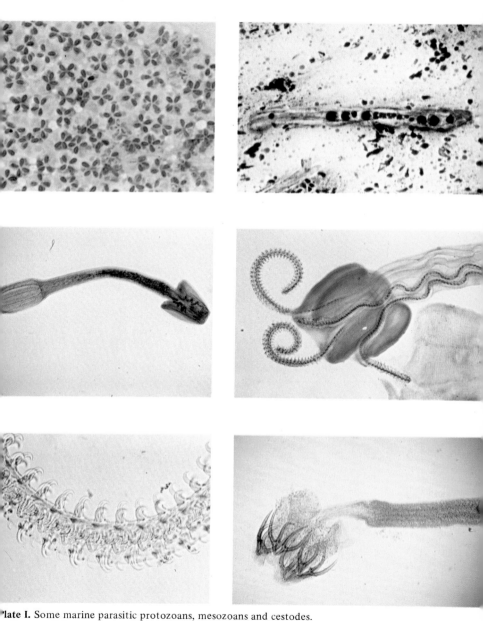

Plate I. Some marine parasitic protozoans, mesozoans and cestodes.
1. Spores of the myxozoan *Kudoa* from the muscles of the kingfish, *Seriola grandis*, on the Great Barrier Reef. **2.** The mesozoan *Dicyema* from the kidneys of *Octopus vulgaris* (specimen courtesy R. B. Short). **3.** Larva of a trypanorhynch cestode from the tuskfish, *Choerodon albigena*, on the Great Barrier Reef; tentacles retracted. **4.** Head of trypanorhynch cestode from the black-tipped rock cod, *Epinephelus fasciatus*, on the Great Barrier Reef; tentacles partly protracted. **5.** Part of a tentacle of (4) more strongly enlarged. **6.** Head of a cestode from the spiral valve of the carpet shark, *Orectolobus ornatus*.

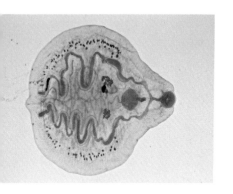 

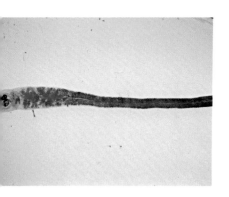 

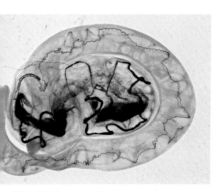 

Plate II. Some marine trematodes.
1. *Staphylorchis* from the body cavity of the graceful shark, *Eulamia amblyrhynchoides*, on the Great Barrier Reef. **2.** *Opecoelina* sp. from the intestine of the bream, *Acanthopagrus australis*, in eastern Australia. **3.** *Nasitrema* from the nasal cavity of a dolphin. **4.** The schistosome *Austrobilharzia terrigalensis* from the blood vessels of a silver gull, *Larus novaehollandiae*, on the Great Barrier Reef; note the narrow female in a groove of the broader male. **5.** The didymozoid *Nemathobothrium* sp. from the mackerel, *Scomber japonicus*, in Argentina. **6.** An immature didymozoid from the fish *Cybiosarda elegans* on the Great Barrier Reef.

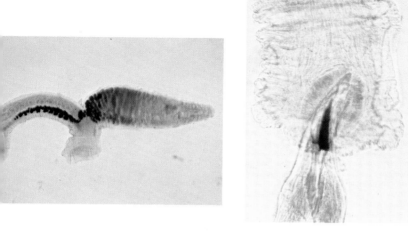

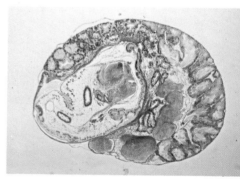
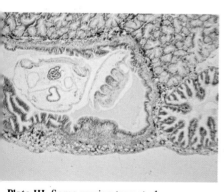

Plate III. Some marine trematodes.
1. *Syncoelium filiferum* from the gill rakers of the skipjack tuna, *Katsuwonus pelamis*, in New Caledonia. **2.** Ventral sucker of *Syncoelium filiferum* attached to a spine of a gill raker. **3.** The aspidogastrid *Rugogaster colliei* from the chimaera, *Hydrolagus colliei*, in the northeastern Pacific (specimen courtesy R. Hobbs). **4.** Section through the digestive gland of the snail *Cerithium monileferum* infected with one aspidogastrid, *Lobatostoma manteri*. **5.** Section through the stomach of the snail *Peristernia australiensis* infected with two *Lobatostoma manteri*; note the thick fibrous sheath beneath stomach epithelium. **6.** Section through stomach of uninfected *Peristernia australiensis*; note the lack of a thick fibrous sheath beneath epithelium.

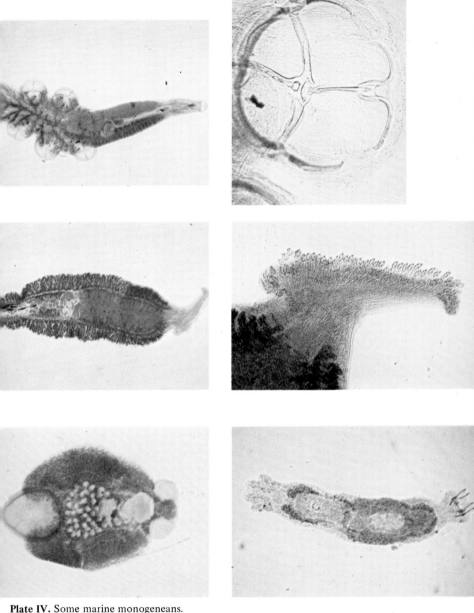

Plate IV. Some marine monogeneans.
1. The polyopisthocotylean *Eurysorchis australis* from the gill arches of warehou, *Seriolella brama*, in New Zealand. **2.** Clamp of *Eurysorchis australis*, more strongly enlarged. **3.** The polyopisthocotylean *Kahawaia truttae* from the gill filaments of the Australian salmon, *Arripis trutta*, in New Zealand. **4.** Posterior attachment organ (opisthaptor) of *Kahawaia truttae*. **5.** A capsalid monogenean (Monopisthocotylea) from the mackerel tuna, *Euthynnus alletteratus*, in eastern Australia. **6.** An ancyrocephaline monogenean (Monoposthicotylea) from the red mullet, *Upeneichthys porosus*, in southern Australia.

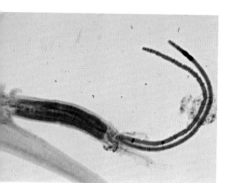

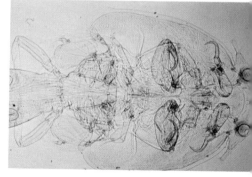

Plate V. Some marine parasitic crustaceans.
1. Praniza larva from the samson fish, *Seriola hippos*, in eastern Australia. **2.** Isopod from the mouth cavity of scad, *Trachurus*, in New Zealand. **3.** The copepod, *Pseudocycnoides armatus*, attached to a gill filament of a broad-barred king mackerel, *Scomberomorus semifasciatus*, in Papua New Guinea. **4.** Head of (3) more strongly enlarged. **5.** The copepod *Caligus* from eastern Australia (photograph F. Roubal). **6.** Ventral view of a male *Caligus katuwo*.

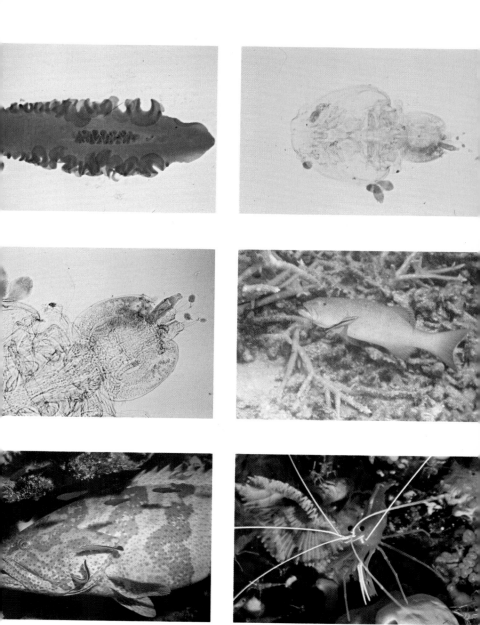

Plate VI. A gyrocotylid and marine hyperparasites and cleaners.
1. *Gyrocotyle fimbriata* from the chimaera, *Hydrolagus colliei* in the northern Pacific (specimen courtesy R. Hobbs). **2.** Hyperparasitic monogeneans, *Udonella caligorum*, and some smaller commensal ciliates attached to the copepod *Caligus brevicaudatus* on the gunard, *Trigla lucerna*, in the North Sea. **3.** Parts of (2) more strongly enlarged. **4.** A cleaner wrasse, *Labroides dimidiatus* partly covered by the operculum of a coral trout, *Plectropomus leopardus*, on the Great Barrier Reef. **5.** A large cod, *Epinephelus tauvina*, with two cleaner fish, *Labroides dimidiatus* (photo W. Deas). **6.** Grabham's cleaner shrimp, *Hippolysmata grabham* (photo W. Deas).

function of preventing infection by parasites, although the function that has been suggested is protection against predation (Winn and Bardach 1959, cited by Hobson 1965).

Examples of cleaning symbioses By far the most common behavioural type of parasite removal among marine animals is cleaning symbiosis, in which one species picks parasites from the body, mouth or gills of another species and eats them. Table 3 gives a list of animal species which have been found as cleaners in freshwater and in the sea. Two species of polychaetes, several species of crustaceans, particularly shrimps, have been observed to be cleaners, but the greatest number of cleaner species is found among the bony fishes. Approximately 60 species of fish belonging to 18 families are known to show cleaning activities, some of them only during their juvenile life. According to Feder (1966):

> Cleaning symbiosis is apparently worldwide in the marine environment, and has been found everywhere in the ocean where there have been extensive diving observations. It seems to involve most fishes as hosts, particularly littoral forms. Reef and pelagic forms, rays, and bony fishes, as well as marine turtles, marine iguanas, and sea urchins are cleaned by specific symbionts.

Apparent occurrence of most cleaner species in warm waters may be due to the fact that relatively more studies have been made in warm oceans because of the better diving conditions.

The cleaner may belong to the same species as the host (Kollatsch 1959) and even birds have been seen to clean marine fish; large ocean sunfish, *Mola mola*, were cleaned by sea gulls while floating on the surface. A young dolphin was observed to nibble at the parasitized skin of another (Gooding and Magnuson, cited by Feder 1966). The Galapagos marine iguana, *Amblyrhynchus cristatus*, while ashore, has ticks removed from its body by a crab, *Grapsus grapsus*, and by finches, *Geospiza* spp.; when submerged, its moulting skin is frequently picked off by the damsel fish, *Abudefduf troschelii* (various authors, see Hobson 1969, Rand 1969). Some examples of cleaning symbioses are discussed in the following.

The polychaete, *Histriobdella homari*, lives in the gill chamber of lobsters, *Homarus americanus* and *H. vulgaris*, and feeds on bacteria, blue-green algae and other microorganisms which grow on the inner surface of the gill chamber, the gill filaments and especially on the surfaces and setae of the skeletal parts between the gills (Jennings and Gelder 1976). Its feeding activities contribute to keeping these vital parts clean.

An example of an association between a shrimp and a nudibranch, which may be interpreted either as a commensal relationship, a

Table 3. Cleaning organisms. Condensed from Feder (1966). Caribbean and Gulf of Mexico listed as part of Atlantic. Some records (one species of polychaete, one species of shrimp, one species of Pomacentridae, two species of Embiotocidae, two species of Syngnathidae, one species of Cichlidae, three species of Labridae, one species of Percidae, five species of Cyprinidae) added from Khalil (1964), Clarke et al. (1967), Gotshall (1967), Hobson (1969), Turner et al. (1969), Abel (1971), Hobson (1971), Potts (1973a), Schuhmacher (1973), Jennings and Gelder (1976) and Sankurathri and Holmes (1976).

Cleaner	Locality
Polychaeta (2 species)	North Atlantic, freshwater
Crustacea Decapoda	
Crab (1 species)	Tropical Pacific (Galapagos Islands)
Shrimps (7 species)	Tropical and subtropical Pacific and Atlantic
Teleostei	
Cyprinodontidae (1 species)	Cold to tropical Atlantic (laboratory observation)
Syngnathidae (3 species)	Mediterranean to North Atlantic (laboratory observation)
Carangidae (2 species, 1 of these juvenile)	One species in tropical and temperate Atlantic and Pacific, one species in tropical Atlantic
Apogonidae (1 species)	Tropical Indian Ocean
Pseudochromidae (1 species)	Tropical Atlantic
Haemulidae (1 species)	Tropical and subtropical Atlantic
Callichthyidae (2 species)	South America (freshwater)
Embiotocidae (5 species)	Cold to subtropical Pacific
Pomacentridae (3 species, 1 of these juvenile)	One species in tropical and subtropical Atlantic, two species in tropical and subtropical Pacific
Labridae (17 species)	Tropical and subtropical Indo-Pacific, two species in Mediterranean, three species in temperate Atlantic (two of these laboratory observations)
Chaetodontidae (4 adult species, juveniles of several species)	Tropical and subtropical Pacific and Atlantic
Scaridae (1 species)	Tropical Pacific
Gobiidae (5 species)	Two species in tropical Atlantic, three species in tropical and subtropical Pacific
Echeneidae (4 species)	Two species in temperate to tropical Atlantic and Pacific, two species in tropical and subtropical Atlantic and Indo-Pacific
Blenniidae (1 species)	Indo-Pacific
Cichlidae (1 species)	Africa (freshwater, laboratory observations)
Percidae (1 species)	Europe (freshwater)
Cyprinidae (5 species)	Europe (freshwater)

permanent cleaning symbiosis or a combination of both, was described by Schuhmacher (1973). This author found four specimens of the nudibranch *Hexabranchus sanguineus* in the course of two years in the Gulf of Aqaba. Three of the molluscs each had one pair of the shrimp *Periclimenes imperator* on their surface. The latter have the same colour pattern as the host, a bright red underground with white markings, which changes to entirely red on a red variety of mollusc. The shrimps are permanently associated with the mollusc and move around on it even when it is swimming. They feed mainly on particles attached to the mucous surface of the snail and on its faeces.

The Pederson cleaning shrimp, *Periclimenes pedersoni*, reaches a length of not more than 4 cm. It is conspicuously striped with white and spotted with violet, but otherwise its body is transparent. It usually lives in association with the sea anemone *Bartholomea annulata* in the western tropical and subtropical Atlantic (Limbaugh *et al.* 1961). Feder (1966) described its cleaning activities as follows:

> Local fishes quickly learn the location of these shrimp and often visit them to be cleaned even if the shrimp has retired into a hole behind the anemone. When the shrimp is available for cleaning, it climbs on an anemone, or a small but prominent rock or coral beside the anemone, and waits for a fish to approach. When a fish enters the area the shrimp whips its long and very conspicuous antennae while it sways its body back and forth. If the fish is responsive it swims to the shrimp and stops 2 to 5 cm away. In general the fish presents its head or gill region; if there is an injured or infected area this part is presented first. When the fish is close enough and quiescent the shrimp will swim or crawl forward and board it. The fish remains almost motionless during this inspection while the shrimp walks vertically over it examining irregularities, tugging at parasites and injured regions. The shrimp makes minor incisions to remove subcutaneous parasites. As the shrimp approaches each gill cover, the fish opens it and allows the shrimp to enter and forage among the gills; it also enters and leaves the mouth cavity. If the fish is alarmed, the shrimp is either forcibly ejected or is given a signal to retreat.

Randall (1962) refers to observations of V. and H. Pederson, according to which *Periclimenes* even makes incisions to remove encysted parasites.

Potts (1973a) described the cleaning activity of the fish, *Crenilabrus melops*, from temperate British waters. There is no elaborate behaviour associated with the cleaning.

> The *C. melops* approaches the plaice and swims above the body of the plaice with its head orientated towards it as if inspecting it. When orientating the *C. melops* hovers by beating of the pectoral fins and

undulating the tail. In this way it could manoeuvre and remove praniza parasites. Not only were parasites removed, but also attempts were made to remove blemishes on the skin and also pigment spots from the epidermis. The plaice showed no obvious cooperation during the occasions that *C. melops* was removing parasites, and often swim away when the wrasse sucked at their skin.

On the Great Barrier Reef, one of the most common Indo-Pacific cleaner fish, *Labroides dimidiatus*, lives in groups of about eight to ten individuals consisting of one large male and a harem of smaller mature females plus a few differently coloured juvenile females. The territory of the group is usually a large coral block. The largest and dominant female shares the whole territory with the male, whereas the smaller females, in hierarchical order, inhabit smaller subterritories within the group territory (Robertson 1973). Off Aldabra, Indian Ocean, small individuals do not appear to have fixed ranges, and large ones occupy more open situations, the adults usually living in pairs (Potts 1973b). Wickler (1968) described the cleaning behaviour of this cleaner fish. When approaching a host fish, it "dances", a behaviour pattern most commonly seen in young individuals (Fricke 1966). The fish swims slowly forwards continuously flipping its abdomen with the spread tailfin upwards, followed by a sinking down of the abdomen to the horizontal. This dance draws attention of host fish to the cleaner (but see the discussion by Gorlick *et al.* 1978 on the function of the dance). As soon as other fish approach, it swims towards them and inspects their surface. While doing so it rapidly beats with its ventral fins and glides over the surface of the host. If the host does not hold still it hits it violently with its wide open mouth.

The cardinal fish, *Siphamia versicolor*, lives in association with the sea urchin, *Diadema*, in the Indian Ocean. Apparently, it cleans the sea urchin and it, in return, seeks protection between its spines when threatened. The close adaptation of the fish to its hosts is indicated by its colour changes. When close to the urchins, the fish has a uniform dark red brown coloration, similar to that of the urchin. When it leaves its host during the night or when it is chased away, the colour pattern becomes silvery with dark brown stripes (Eibl-Eibesfeldt 1961).

McCutcheon and McCutcheon (1964) established a cleaning relationship between the Black Sea bass *Centropristes striatus*, a predatory fish, and topminnows, *Fundulus heteroclytus*, in an aquarium. Both species live together along the north American coast in temperate waters. The bass was experimentally infected with parasitic copepods, *Lernaeenicus radiatus*. Some topminnows were eaten more or less immediately, but others cleaned the bass which seemed to be paralysed or insensitive during the cleaning, mutuality and cooperation being shown by the fact that the bass exposed vulnerable gill regions. Copepods were eaten by the topminnows, and all topminnows were eaten by the bass before the fourth day.

Morphological adaptations and guild signs of cleaners Some cleaner fish have distinct morphological adaptations to cleaning. The dentition of the cleaner fish *Symphodus melanocercus*, for instance, differs distinctly from that of non-cleaning species of the same genus. The anterior row of teeth is well adapted to picking up parasites, up to five anterior teeth being in close contact, whereas in the other species only two are in contact. Furthermore, the upper lip of the cleaner is more strongly reduced than in other labrids (Casimir 1969).

Many cleaning animals, shrimps as well as fish, particularly in the tropics, have conspicuous "signal" markings which are often remarkably similar. Eibl-Eibesfeldt (1959), for instance, drew attention to the similar colour patterns of the goby, *Elacatinus oceanops*, a cleaner fish in the Caribbean, and the wrasse, *Labroides dimidiatus*, a cleaner fish in the Indo-Pacific. He suggested that such similar conspicuous markings are "guild signs", signs which enable host fish to distinguish cleaner organisms from non-cleaners. Potts (1968, 1973a) gave further examples. The Californian señoritas, *Oxyjulis californica*, and the Mediterranean *Crenilabrus melanocercus* are both small wrasses with a black spot on the tail or caudal peduncle, and the British species *C. rupestris* and *C. exoletus* are both very similar to these species. According to Feder (1966), even the colour pattern, strong longitudinal striping, of the shrimp *Hippolysmata grabhami* (although not the actual colours) is very similar to that of the cleaner fishes *Labroides dimidiatus* and *Elacatinus oceanops*.

Cleaning stations Eibl-Eibesfeldt (1955), Randall (1958, 1962) and many authors after them reported that cleaner fish are often found around so-called cleaning stations to which host fish come to be cleaned. According to Limbaugh (1961) "the various species of cleaning fish and shrimp tend to cluster in particular ecological situations; at coral heads, depressions in the bottom, ship wreckage or the edge of kelp beds". The number of host fish cleaned at such stations may be considerable. Limbaugh saw up to 300 fish cleaned at one small station in the Bahamas during one 6-hour daylight period. Some of the fishes passed from station to station and returned many times during the day, and those that could be identified easily returned day after day at regular intervals. Limbaugh reports that it seemed that many of the fishes spent as much time at cleaning stations

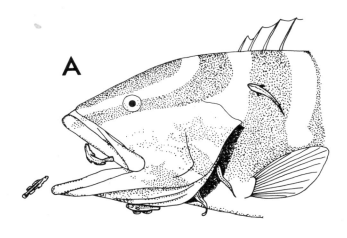

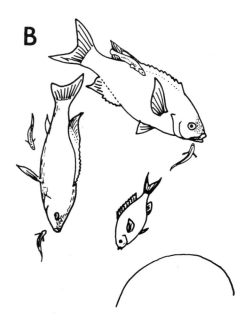

Fig. 59. A. Several *Elacatinus oceanops* cleaning the head of an *Epinephelus*. B. Cleaning postures of two *Clepticus parrae* and one *Chromis marginatus* cleaned by juvenile *Thalassoma bifasciatum*. Redrawn after Eibl-Eibesfeldt (1955).

as they did in feeding. Feder (1966) quoted work by Gooding and Magnuson (1964) and Gooding (1964) which indicates that drifting oceanic objects may be cleaning stations for pelagic fish. Pelagic fish were also seen to visit the benthic cleaner, *Crenilabrus melanocercus* (Potts 1968). Slobodkin and Fishelson (1974) even claimed that the cleaning stations of the cleaning wrasse at Eilat in the Red Sea represent the same kind of focal location for animal aggregation as do water holes, animal kills or garbage dumps (see discussion in Gorlick et al. 1978). However, cleaning stations are not established by all cleaner fish. According to Hobson (1969), for instance, the señorita (*Oxyjulis californica*) and the sharp-nose seaperch (*Phanerodon atripes*), known cleaners of inshore species of southern Californian fishes, do not establish well defined stations.

Host and site specificity of cleaners   A certain species or individual of cleaner fish may clean a wide variety of hosts, but in some cases a tendency towards host and site specificity could be demonstrated. Examples of wide host range were given by Youngbluth (1968), who observed that the cleaner fish, *Labroides phthirophagus*, cleaned approximately 50 species of host fish belonging to 20 families in the field, and by Potts (1968), according to whom the labrid, *Crenilabrus melanocercus*, approached 17 fish species, of which 12 benthic and pelagic species responded by adopting an invitation posture and were cleaned. Tendency towards host specificity was shown by Hobson (1971) who observed that, although a considerable number of fish species were cleaned by the señorita in Californian waters, only few species were cleaned frequently, and although señoritas as a group clean a number of different fishes, a given individual tends to initiate cleaning with members of one species only. Potts (1973a) found that there is not only a specificity for certain host fish, but that there is also a preference for certain parts of the host fish which are cleaned. The cleaner, *Crenilabrus melops*, cleaned the pelvic fin region and vent of host fish 57 per cent of the time, the sides of the body 17 per cent and the head 11 per cent. The frequency of cleaning did not relate to the distribution of parasites on the various parts of the body. *Crenilabrus melanocercus* cleaned mostly the base of the fins (Potts 1968).

Although host fish are usually larger than their cleaners, it is not rare to see species of *Labroides* picking at hosts smaller than themselves (Randall 1958). Nocturnal fishes also are frequently cleaned by various species of *Labroides* during the day (Gorlick et al. 1978).

Host behaviour to cleaners   Cleaning is sometimes initiated by the cleaner animal, sometimes by the host fish (Potts 1968, Losey 1971). Randall (1958, 1962) and Pederson (quoted by Feder 1966) observed that fishes may come from long distances to sites occupied by the cleaners and not just from the immediate community. Fish often assume typical poses (Fig. 59) and they may also change their colour while being cleaned (Feder 1966). Feder reports that this colour change is often dramatic and the colour is often in contrast to the environment. "Black-hued surgeon fish (*Acanthurus achilles*) will turn bright blue when picked over by *Labroides dimidiatus*. As a gold fish (*Parupeneus trifasciatus*) approaches a *L. dimidiatus*, the former may change its colour from light tan to moderate pink (Randall 1958)." Hobson (1971) described in detail the interactions of host fish with the señorita, *Oxyjulis californica*, their cleaner fish. Cleaning in the blacksmith, *Chromis punctipinnis*, is usually initiated by a señorita swimming up alongside a blacksmith in mid-water and closely inspecting its body.

> The blacksmith may then immediately stop swimming and, holding its fins motionless and erect, drift into an awkward-appearing posture. Usually the blacksmith is head down, but sometimes turns on its sides or its tail down. On some occasions the blacksmith presents a particular part of its body to the inspecting señorita. The señorita swims about this fish, usually pausing briefly to pick at its body. Immediately following the first sign of this activity, other blacksmiths converge on the spot, so that very quickly 10 or more crowd around the cleaner.

It is very rare that blacksmiths solicit cleaning from a señorita that did not make the initiating approach. In the cleaning of the topsmelt (*Atherinops affinis*) the señorita swims up to an individual and inspects it. Immediately other topsmelt converge and solicit attention. Topsmelt frequently have their tail down in contrast to the head down posture often assumed by blacksmith. If a garibaldi, *Hypsypops rubicunda*, is cleaned the señorita swims up to it and closely inspects its body. Usually the garibaldi hovers motionless in a normal horizontal attitude, its fins sometimes erect. Halfmoons (*Medialuna californienis*) are cleaned like a blacksmith if many individuals are present, but like a garibaldi if only a single individual is present. Señoritas may also clean members of their own species. Usually a single individual is attended which hovers motionless in a normal horizontal attitude, except that its fins are erected. Sometimes the mouth is opened wide and gill covers are distended.

Wahlert and Wahlert (1962) noticed that host fish visiting a cleaning station of the cleaner fish, *Crenilabrus melanocercus*, in the western Mediterranean assume their invitation posture before the cleaner fish shows itself, and according to Casimir (1969), the cleaner fish, *Symphodus melanocercus*, in the Mediterranean can distinguish different intensities of the invitation posture of host fish. Given the choice, it prefers the more intense signal. Some host fish are attracted to the cleaning territory of the cleaner fish, *Labroides dimidiatus*, at Aldabra, Indian Ocean, and may adopt invitation postures even in the absence of cleaners (Potts 1973b).

The 12 host fish of the cleaner fish, *Crenilabrus melanocercus*, have a common invitation posture, which consists of erecting all unpaired and sometimes also the paired fins, with the mouth open and extended forwards and the opercular covers raised. The fish drift in this position, in some species with the head up, in others with the head down; some rest on one side or do not give any obvious invitation (Potts 1968). Losey (1979) showed that the presence of ectoparasites had little effect on the host responses to cleaner fish in *Chaetodon auriga* and only amplified the responses in *Zebrasoma flavescens*.

Invitation postures often are assumed close to the edges of softly waving gorgonians, where cleaners are never found. Such postures may last for considerable periods and the same gorgonian may be visited repeatedly by the same fish coming over a long distance (Thresher 1977). Such behaviour has been called pseudocleaning (Thresher 1977) and is in agreement with the suggestion by Losey (1972b, 1975) and Losey and Margules (1974) that the proximate causation of cleaning is reinforcement by tactile stimulation, normally provided by the cleaner, but in this case provided by the waving gorgonian. Cleaning postures also can be experimentally induced by artificial tactile stimulation (Losey 1976, 1977).

Before fishes leave the cleaner, they often signal that they are ready to leave. Similar signals are given when danger forces the host fish to leave. Eibl-Eibesfeldt described that a grouper being cleaned by the cleaner fish, *Elacatinus*, invariably performs a "ritualised intention-movement of shutting its mouth" which causes the cleaner fish to scuttle out and leave. *Epinephelus striatus* cleaned by neon gobies holds its mouth and gill slits rigidly open with intervals for breathing. When it wishes to terminate cleaning, the host closes its mouth sharply but not completely, and then opens it widely again. This signal indicates to the goby it should leave. The same sign is given when the host fish is frightened (Eibl-Eibesfeldt 1955, quoted by Feder 1966). Randall (1962) described the signal given to the cleaner, *Labroides dimidiatus*, by a moray eel. The eel jerked its head to one side when it wished to terminate cleaning, and the cleaner swam rapidly out of its mouth. Potts (1968) reported that after *Serranus scriba* had finished being cleaned by the cleaning fish, *Crenilabrus melanocercus*, it ceased to posture and quickly turned towards the cleaner as if to chase it. This caused the cleaner to swim away. Such signals were not observed in any of the other 11 species cleaned by this cleaner. Gorlick *et al.* (1978) doubted that such host signals have the "altruistic function suggested above. It is quite possible that these host behaviours are in response to painful stimulus . . . it may also represent conflict between the tendencies to approach (or remain posing) and withdraw from the cleaner."

Randall (1958) raised the question whether host fishes learn the role of the cleaner *Labroides* through experience or whether their behaviour is innate. He observed that parrot fish which had been in an aquarium for at least several months followed a *Labroides* which was added to the aquarium, "stopping occasionally to literally stand on their tails, waggle their pectoral fins, and seemingly to ogle the labrid, as if to entreat it to perform the expected services", an observation which indicates that the behaviour is either innate or has been remembered for a long period. Herald (1964) observed that fish from geographical areas where the cleaner fish *Labroides* does not occur, reacted immediately to this fish in an aquarium, indicative of an inborn response.

In well established cleaning symbioses, cleaners are rarely eaten by their hosts. Some exceptions are described by Strasburg (1964), who found specimens of three species of Echeneidae in the stomachs of several species of large predatory fish, but he did not exclude that the sucker fish had been dead prior to being eaten or that they were dislodged from their normal habitat, the buccopharyngeal region, when swallowed.

Parasite removal by cleaners A number of observations show that cleaners remove ectoparasites. Szidat and Nani (1951) were apparently the first to give evidence. They found ectoparasitic caligoid copepods in the stomach of *Remora remora*, a cleaner fish belonging to the Echeneidae. Maul (1956) found a caligoid as well as an oxycephalid amphipod in the stomach of the same fish species (see also Strasburg 1959, and Cressey and Lachner 1970). Direct observations were made by Potts (1973a), who saw that the pipefish, *Entelurus aequoreus, Syngnathus typhle* and *S. acus*, removed caligoid copepods and praniza larvae, as well as white blemishes, from host fish. Potts also observed that the cleaner fish, *Crenilabrus melops*, removed parasitic praniza larvae of the isopod, *Gnathia maxillaris*, from the bodies of host fish kept in an aquarium. The parasites were removed by a sucking action which appeared to cause some discomfort to host fish. The same cleaner fish was seen to remove praniza parasites from the skin of plaice in an aquarium. It also made attempts to remove skin blemishes and epidermal pigment spots. Mackerel (*Scomber scrombrus*), while harbouring many praniza larvae, were cleaned only rarely because the host fish usually kept out of range. Observations in the natural environment confirmed those in the aquarium. Praniza larvae were found by Potts among the stomach contents of *Crenilabrus melops* caught in the sea. However, the author did not rule out that these parasites may have been picked up from the bottom substrate. The same author observed that the related cleaner fish, *Ctenolabrus rupestris*, removed parasites, mainly praniza larvae from fish in an aquarium, and Wilson (1962, *cited* by Potts 1973a) described how the pipefish, *Entelurus aequoreus*, removed crustacean parasites from the John Dory (*Zeus faber*). Youngbluth (1968) examined the gut contents of the cleaner fish, *Labroides phthirophagus*. He found that the diet both of juveniles and adults consisted primarily of parasitic caligoid copepods, and to a lesser degree of parasitic lernaeid copepods and larval gnathiid isopods. He did not find free-living crustaceans in the intestine, although data from other authors (see Randall 1958) suggest that the two species of cleaner fish, *Labroides phthirophagus*, and *L. dimidiatus*, occasionally feed on free-living crustaceans. To test the possibility of ingestion of free-living crustaceans, Youngbluth kept three cleaning fish in separate aquaria and starved them for three, four and five days, respectively. Zooplankton with a conspicuous proportion of copepods, nauplius larvae and polychaetes was added. The fish were killed the following day and their intestines found to be empty.

Monogenean ectoparasites have not been found in the gut contents of cleaners, probably because their soft body is rapidly digested, and because of the minute size of their hooks and other hard skeletal elements. But there is other evidence for consumption of Monogenea by cleaner fish. Khalil (1964) observed that the small monogenean, *Macrogyrodactylus polypteri*, on the skin of the African lungfish, *Polypterus senegalus*, increases rapidly in numbers (several hundreds to several thousands) if fish are kept in the aquarium, but that hardly any worms are left a few hours after specimens of the cichlid, *Tilapia nilotica*, are introduced. He observed that *Tilapia* picks the worms one by one from the surface and eats them. Kearn (1978) noticed that two of three monogeneans on the skin of a shovel-nose ray were cleanly severed transversely just anterior to the opisthaptor, probably by leather jackets kept in the same aquarium.

Atkins and Gorlick (1976) demonstrated that there were slight, but statistically significant, changes in the population size of a species of ectoparasite on one host species after removal of all *Labroides dimidiatus* from patch reefs at Enewetok Atoll.

Some cleaner fish feed mainly on free-living animals, and

ectoparasites represent only a small proportion of their food. Hobson (1971) showed this, for instance, for the señorita, *Oxyjulis californica*. In 26 specimens of this fish he found the following food items in their stomachs (ranked as percentage of the entire sample): bryozoans encrusted on algae, 43%; caprellid amphipods, 32%; fish eggs, 3%; gammarid amphipods, 2.5%; unidentified crustacean fragments, 4%; and pelecypod molluscs, 2.4%. Items of less than 1 % of the sample included crab fragments, gastropods, pycnogonids and a single gnathiid isopod larva. Unidentified material constituted 16 % of the sample. The isopod larva was the only evidence of ectoparasites.

Latitudinal differences in cleaning symbiosis According to Feder (1966), who based his conclusions largely on Limbaugh's (1961) observations, cleaner fish from colder water are usually highly gregarious or schooling, whereas those species from warm water are solitary, paired or slightly gregarious. Known cleaners of temperate waters seem to be more numerous as individuals than those of the tropics, though the number of species is less. The tropical species are more nearly full time cleaners, and receive a larger proportion of their food in this way. Cool water cleaners simply approach a host fish to clean it, or a cleaner may be approached by the hosts, whereas tropical species very often have elaborate displays, similar to the mating of male fishes, and cleaners in tropical waters are more brightly coloured and more contrastingly marked.

Hobson (1969) critically reviewed these generalizations on the basis of his experience in warm temperate waters of California and in the tropical waters between the Gulf of California and the Galapagos Islands, the same area in which Limbaugh made most of his observations. He arrived at the following conclusions. Firstly, many tropical cleaners are highly gregarious. For instance, cleaning stations of *Heniochus nigrirostris* often have several hundred individuals, and *Abudefduf troschelii* in the tropics is also highly gregarious; some cleaners from California, on the other hand, are solitary or occur in small groups. Furthermore, many cleaners in the tropical eastern Pacific only infrequently engage in cleaning activities, and although one might expect that there would be some more highly specialized cleaners in the tropics, because of the greater number of tropical species, this does not permit the statement that tropical cleaners are generally more specialized. No cleaner in the eastern Pacific performs elaborate displays like species of the genus *Labroides* for instance, and there is therefore no basis for attributing such activity to tropical cleaners in general. Finally, many tropical cleaner fishes are indeed more brightly and more contrastingly coloured than cleaner fish from colder waters, but this has to be expected because fishes in the tropics tend generally to be more colourful. According to Hobson, brightly coloured cleaners belong to groups which also include brightly coloured non-cleaner species. Evidence supporting a direct relationship between bright coloration and cleaning is still weak (but see the section on "guild signs" of cleaners, page 77). Furthermore, several California cleaner species are highly conspicuous. Hobson concluded that generalizations seem to be unwarranted, and more data are needed from the cold temperate and arctic regions.

Mimics The ancient nature of cleaning symbiosis is indicated by the existence of cleaner mimics. *Aspidontus taeniatus* (family Blenniidae) has a colour pattern and a dancing behaviour like that of the cleaner fish, *Labroides dimidiatus* (family Labridae). It approaches other fish in the manner of a cleaner, and fish which have not learned to distinguish the mimic from the cleaner are suddenly attacked and little pieces punched out of their fins (Eibl-Eibesfeldt 1959, Wickler 1960, 1961).

Ecological function of cleaning symbiosis In spite of a considerable number of studies on cleaning symbiosis, little is known about its ecological function. Limbaugh (1961) removed all the known cleaning organisms from two small, isolated reefs in the Bahamas where fish seemed particularly abundant. Within a few days the number of fish was drastically reduced; within two weeks almost all except the territorial fishes had disappeared. Many of the fish remaining developed fuzzy white blotches, swellings, ulcerated sores and frayed fins. However, the experiment was not quantified nor controlled, although the observed contrast of the fish populations of nearby coral heads was, according to the author, very striking.

Youngbluth (1968) could not verify the observations of Limbaugh. He compared three reefs: from one all the cleaner fish of the species *Labroides phthirophagus* were removed, from the second all but one cleaner were removed, and the third was left undisturbed. One month after depopulation the rate of feeding by cleaner fish on the partially depopulated reef did not differ significantly. On the reef without any cleaners, the density of host fish did not appear to be changed, and no serious infection of host fish was noticed. It should be noted, however, that the time of one month may have been too short to permit a significant effect on parasite numbers. Losey (1972a) working on the same cleaner fish as Youngbluth, found that, contrary to the latter author, removal of most of the cleaner fish from a reef resulted in an increase in cleaning behaviour by the remaining cleaners and in changes in the distribution of the host fish. He did not find an increase in the level of infection by ectoparasites of the host fish after all the cleaners were removed, as compared with a similar control reef. Again, it must be stated that the time between removal of the cleaner fish and assessment of parasite infection was less than one year, possibly too short to allow a significant increase in parasite numbers. Atkins and Gorlick (1976) observed a slight but significant increase in population size of an ectoparasitic species after removal of all *Labroides dimidiatus* from patch reefs at Enewetok Atoll.

Wahlert and Wahlert (1962) drew attention to the fact that North Sea fish have generally more ectoparasitic caligoid copepods than fish in the Mediterranean; furthermore, whereas the Mediterranean caligoids are usually opaque and difficult to notice on their hosts, the species in the North Sea are strongly coloured and conspicuous on the silvery bodies of the fish. This may indicate greater cleaning activities in the Mediterranean which leads to the survival of less conspicuous ectoparasites only.

In summary, it is clear that cleaner fish do indeed remove parasites of host fish. It is probable that some of these cleaners are predominantly or even obligatory cleaners, and feed very little or not at all on free-living organisms. Evidence is conflicting, however, on the ecological importance of cleaning activities for host populations, although the widespread occurrence of cleaning symbioses, the cooperation of host fish and the existence of cleaner mimics suggest that some symbioses must have great ecological importance and a long evolutionary history.

Immunity and tissue reactions

Immunity and tissue reactions in vertebrates All defence reactions of the host at the humoral and tissue level are similar in one respect, they are based on the ability of the host to distinguish self (its own cells) from non-self (foreign cells and material). Vertebrates have three kinds of defence mechanisms, phagocytosis, inflammation and adaptive immunity, of which the first two are non-specific tissue reactions to non-self or tissue damage, and the third is specific to a certain type of non-self material. Smyth (1962) and Wakelin (1976) reviewed these reactions to parasite invasions, and many detailed examples of host responses to various parasite infections were given by Pflugfelder (1977). The reactions, and particularly immunity reactions, are best developed in birds and mammals, but are also found in the "lower" vertebrates. Corbel (1975) gave a recent review of the immune response in fish, with an extensive bibliography. All three defence mechanisms usually interact and occur in most body tissues and organs. They are weakest or altogether absent in sites with few or no fibroblasts, for instance in the eye chamber or parts of the central nervous system. Typically, tissue reactions to a foreign invader occur in a certain sequence. Degeneration or necrosis of cells due to the infection leads to an inflammatory reaction characterized by an oedema (swelling of tissue) as a result of the accumulation of white blood cells and plasma and the vasodilation (widening) of blood vessels. Certain cells of the host phagocytoze, that is, engulf and break down small parasites. If the parasite is not eliminated, a chronic or long lasting inflammation develops, characterized by formation of a connective tissue capsule around the parasite. Macrophages (large phagocytic cells) in the capsule phagocytoze damaged cells and finally the parasite.

Special defence mechanisms may be active on the body surface. For example, fish continually shed a mucoid material from the skin even if uninfected, but the density of the slough increases when monogeneans are present and the parasites are removed with the slough (Lester 1972, Lester and Adams 1974).

Immunity reactions are due to certain molecules of non-self material called antigens which induce the host to form antibodies. Each antibody corresponds to a certain antigen, and is specific for it. In an antigen–antibody reaction, the two combine and the foreign antigen is often neutralized by the antibody, the host thereby acquiring an immunity. However, not all antibodies give such protection. Those which do are the so-called protective antibodies, formed as a reaction to the functional or essential antigens. There is evidence that fish may be induced to produce antibodies to monogeneans which appear in the surface mucus (Fletcher and Grant 1969, Di Conza and Halliday 1971), and to acquire an immunity (Jahn and Kuhn 1932, Nigrelli and Breder 1934, Nigrelli 1937).

Parasites have developed various and sometimes ingenious ways of evading the host responses or even of utilizing them for themselves. Thus, tissues formed as a reaction to an invasion may be used as food by the parasite, and the defence reactions may so weaken the host that it becomes easy prey to a predator which may be important for completion of the parasite's life cycle (Wakelin 1976). Bloom (1979) has reviewed the mechanisms whereby parasites evade the immune response. One striking method is acquisition by parasites of host antigens on their surface, which makes the host respond as though dealing with "self-material". Certain parasites have developed the ability to live in phagocytes which normally have the function of ingesting and digesting invaders, and in some extreme cases the parasites actually need part of the host's immune system for survival. Such mechanisms have not been studied in marine parasites.

Tissue reactions in invertebrates In invertebrates, tissue reactions are mainly responsible for defence against parasites, although in some cases body fluids are involved. Thus, Tripp (1966) demonstrated the presence of a protein which agglutinates red blood cells, in the blood of the American oyster, *Crassostrea virginica*. Wakelin (1976) concluded that there is little evidence that the factors responsible are induced by the non-self material, and there is no evidence for enhanced production of protective factors at subsequent invasions. Altogether, there appears to be consensus among biologists that invertebrates are not capable of producing true antibodies (Cheng 1975a). Although the oyster, *Crassostrea virginica*, parasitized by larvae of the trematode *Bucephalus* sp. or the protozoan *Minchinia nelsoni* and the snail *Biomphalaria glabrata*, parasitized by schistosomes, show increases in certain fractions of serum protein (Feng and Canzonier 1970, Gress and Cheng 1973), it is doubtful that these represent immunoglobulins, the proteins corresponding to those which have antibody function in vertebrates (see also Bang 1970 for marine arthropods). But even though formation of antibodies due to a previous infection has not been demonstrated in invertebrates, an acquisition of resistance has been postulated which is not due to selective survival of more resistant individuals. Andrews (1968) postulated this because an early exposure of oysters to the ascetosporan, *Minchinia nelsoni*, appeared to be important for the subsequent survival of large oysters.

In molluscs, where defence reactions to parasites are best known, phagocytosis by haemolymph cells appears to be the major defence mechanism against invading microorganisms (Cheng 1963b, Cheng 1975a). Tripp (1958) demonstrated phagocytosis of infected particulate matter in the oyster, *Crassostrea virginica*. Intracellular digestion and migration of the phagocytes from the blood channels to the tissues and finally through the epithelium to the exterior disposes of the foreign material. Tripp (1963) distinguished the following possibilities in molluscs: (1) There is no marked cellular response; (2) small foreign particles in the tissues and blood sinuses are phagocytozed and either carried through the epithelium to the exterior or degraded within the cells; (3) foreign material is infiltrated and surrounded by amoebocytes and the lesion is eventually walled off by an epithelial layer (for an early study on phagocytosis and ejection of foreign material see Takatsuki 1934; for a review of cellular reactions in marine molluscs to helminths, see Cheng and Rifkin 1970).

The process of inflammation is remarkably similar to that in vertebrates. Pauley and Sparks (1965) described it in the Pacific oyster, *Crassostrea gigas*, elicited by turpentine injection. Areas of pus could be seen without a microscope after 40 hours. Histological examination revealed oedema, leucocytic infiltration and congestion of the small blood channels after 8 hours. By 16 hours the vessels had become markedly dilated and the larger vessel walls were paved with leucocytes. At 24 hours, there was a cellular exudate and migration of leucocytes toward the source of irritation, and at 48 hours the wound was surrounded by a thick layer of leucocytes. After 64 hours, multinucleate giant cells began to appear and large numbers of leucocytes migrated across the epithelium of the digestive system. At 72 hours, much pus in the area of the wound was apparent, and oedema, strong infiltration with leucocytes and distention of the gonadal ducts were visible in the gonads, with leucocytes forming walls around the necrotic areas. At 88 hours, distinct abscesses were visible, characterized by liquefaction necrosis, surrounded by necrotic leucocytes and peripheral normal leucocytes.

Capsule formation in molluscs may be a very complex process, as shown for instance by Yoshino (1976) in the marine snail, *Cerithidea californica*, infected with larval trematodes of *Renicola buchanani*. Special blood cells without granules called hyalinocytes are responsible, and three phases of encapsulation could be distinguished. In the first phase, hyalinocytes aggregate around each sporocyst, in the second phase, many hyalinocytes form a dense network of cells close to the parasite's surface, and in the third phase, a capsule four to eight cell layers deep and tightly adhering to the parasite is formed by horizontal flattening of the hyalinocytes. The sporocysts are not harmed. Cheng and Rifkin (1970) distinguished five types of encapsulation, characterized by different constituents and processes, and a variety of reactive elements. There are various types of blood cells, several types of connective tissue cells and fibres, three types of muscle fibres, and so-called "brown cells". Cellular reactions, according to the same authors, are phagocytosis, encapsulation, leucocytosis (increase in number of blood cells) and "nacrezation" (pearl formation by deposition of nacre around a parasite in the mantle region of the mollusc). Pearl formation induced by trematode and cestode larvae has been described frequently (Dubois 1901, Giard 1903, 1907, Jameson 1912, Southwell 1924, review by Stunkard and Uzmann 1958). According to Southwell (1924), cestode larvae induce it in the Ceylon pearl oyster only when the parasite dies before a cyst is formed and induces a local irritation. Other bodies, such as sand grains, can do the same.

The strength and type of tissue reaction depend on the site of infection. Thus, larval cestodes, *Tylocephalum*, induce capsule formation by the host, but the thickness of the capsule in the zone beneath the host's digestive tract and in the digestive gland is different, apparently depending on the abundance of Leydig's tissue (Rifkin and Cheng 1968), and according to Cheng *et al.* (1966) the fibrous constituent of the fibrous capsule surrounding larval trematodes, *Himasthla quissetensis*, in the tissues of eight species of mussels, appeared to depend on the type of fibres available in the immediate proximity. The aspidogastrid, *Lobatostoma manteri*, usually lives in the stomach of the marine snail *Peristernia australiensis*, where it induces the formation of a thick layer of fibrous tissue around the stomach, and in the digestive gland of the snail *Cerithium moniliferum*, where it induces proliferation of connective tissue in the gland (Rohde and Sandland 1973).

The type of reaction may also depend on the state of the parasite. Healthy parasites often are only slightly encapsulated, whereas dead ones which are being resorbed by the host tissue are surrounded by thick capsules of fibres and leucocytes (Cheng 1966b). Resorption of dead parasites was described by Cheng and Rifkin (1968) in the clam, *Tapes semidecussata*, infected with larvae of the cestode *Tylocephalum*. Living parasites are encapsulated by fibres and leucocytes between the digestive diverticula. Resorption is initiated by mass migration of leucocytes and so-called "brown cells" which form a cellular wall around the parasite which now disintegrates. Disintegration of the capsule follows next, and the cellular remainder of the parasite is phagocytozed by leucocytes.

Effects on host individuals

Effects on invertebrates There are numerous studies on how parasites affect their invertebrate hosts, and only a few examples will be discussed here. For further examples see Pflugfelder (1977), who gives many details from various groups of parasites and hosts, and Cheng (1967) who discusses parasites of molluscs.

Effects on the host are usually complex and those stressed by the respective authors are always the result of the methods used. For example, histological and histochemical methods have revealed mechanical damage as well as various types of cell and tissue damage, and changes in tissue types and chemical composition, whereas physiological methods revealed changes in growth rate, resistance to thermal stress, changes in assimilation and ingestion. Observations of the behaviour of animals have shown that infection may significantly alter behaviour patterns.

The complexity of effects was, for instance, shown by the study of James and Bowers (1967a) of *Cercaria bucephalopsis haimaena* in the bivalve, *Cerastoderma edule*. The primary effect apparently is a mechanical compression of digestive ducts and tubules which cut off more distal tubule cells from their food supply. This leads to starvation autolysis. The cut off cells have a reduced content of glucose, glycogen, glycoprotein, phospholipid and proteolipid food storage globules and acid mucopolysaccharid. Neutral lipids, fatty acids, alkaline phosphatase, acid phosphatase and non-specific esterase are increased, and there is also a compensatory increase in the number of food vacuoles. In the visceral haemocoel near the parasites, fatty acids, neutral lipids and acid mucopolysaccharide are increased and glycogen is reduced. James and Bowers (1967b) showed that the parasites absorb the host's glucose, glycogen, fatty acids and neutral lipids. In the case of the bivalve *Musculium partumeium* infected with the trematode *Gorgodera amplicava*, the primary effect also appears to be mechanical in the form of a distension of the lamellae and occasional ruptures. As a result of this, necrosis of the mechanically severed gill epithelium occurs (Cheng 1963a).

Some other examples of effects by parasites on molluscs are given in the following. The copepod, *Mytilicola orientalis*, induces a metaplasia in the gut of the oyster, *Crassostrea gigas*, and of the mussel, *Mytilus edulis*. Whereas the normal gut epithelium is tall and columnar, it is low and cuboidal or squamous in infected hosts. The mucosa is sometimes completely destroyed, parts of the parasite penetrating into the underlying connective tissue, where a tendency towards fibrosis exists (Sparks 1962, Moore *et al.* 1978). In *Littorina littorea*, trematode larvae destroy the tissue of the digestive gland and the released yellow–orange pigment stains the foot, which facilitates recognition of infected snails (Willey and Gross 1957). The mechanism of the effect on oysters, *Crassostrea virginica*, by the larval trematode, *Bucephalus*, was studied by Cheng and Burton (1966) using histochemical methods. Whereas glycogen in uninfected oysters is stored in various tissues, there is a marked reduction in stored glycogen in infected oysters. In the latter, demonstrable amounts of glucose are found in the blood vessels and sinuses, apparently because of the destruction of the sites of glycogen synthesis and storage. Effects by trematode parasites on the snail *Littorina littorea* are, according to Watts (1971), an increase of free amino acids in the head–foot muscle by 10.9 per cent in infections with rediae of *Cryptocotyle lingua*, a decrease of 12.7 per cent in infections with rediae of *Himasthla leptosoma*, and a decrease by 57.5 per cent in infections with sporocysts of *Cercaria emasculans*. The crab *Fabia subquadrata* causes extensive damage to its bivalve hosts by eroding their gill filaments and causing blisters of the mantle (Pearce 1966). In the snail *Turritella* infected with sporocysts of *Cercaria doricha*, Negus (1968) found no trace of toxic or mechanical damage, and there was no depletion of amino acids or purines, but the concentration of plasma proteins was higher in parasitized snails and the concentration of free amino nitrogen in the head–foot muscle showed an increase of 7 per cent. There was no removal of glycogen by the sporocysts, but there was an inverse relationship between the number of sporocysts and the amount of gonadal tissue.

The various tissue and functional changes may lead to a generally reduced "performance" of the hosts. Thus, snails of the species *Nassarius obsoleta* heavily infected with trematode larvae are less resistant to temperature stress than uninfected ones, and there may be a reduced resistance to a low oxygen level (Vernberg and Vernberg 1967). There also may be a depressed physiological state as indicated by a reduced weight. Thus, the copepod, *Mytilicola orientalis*, lowers the condition of the oyster, *Ostrea lucida*, as measured by its volume (with or without shell) (Odlaug 1946, Cole and Savage 1951) or the weight of the meat (Mann 1951). According to Meyer and Mann (1950), the damage is due to an acceleration of protein digestion resulting in increased use of oxygen, at the same time leading to a decrease in filtering ability. The combined effect is deficient growth of the host. Weight per unit of volume of the shell cavity in *Crassostrea virginica* is reduced by the crab *Pinnotheres ostreum* (see Haven 1959). According to Stauber (1944), crabs injure the oyster's gill filaments which interferes with its normal feeding. In extreme cases, the result of infection may be death. For example, the oyster *Ostrea lutaria* under experimental conditions had a much higher death rate when infected with an unidentified trematode than when uninfected (Millar 1963).

Phillips and Cannon (1978) and Bishop and Cannon (1979) observed complex changes in the behaviour of sand crabs, *Portunus pelagicus*, parasitized by the barnacle, *Sacculina granifera*. With regard to alarm reactions, feeding and locomotion, the infected crabs behave as normal crabs, but general body care is increased as a result of the increased epizoic fauna on the exoskeleton which develops because the parasite inhibits moulting. Furthermore, infected crabs with an external "sac" of the parasite resemble egg-carrying females in terms of their moderate daylight activity and particularly the grooming care and attention given to the sac. Parasitized crabs excavate a depression and using the third walking legs (and often second and first as well) attend the sac as an egg-carrying female attends the egg mass.

A common phenomenon, known from a number of host and parasite groups, is parasitic castration, or sterility due to parasite infection. Reinhard (1956) reviewed parasitic castration in the Crustacea and Baudoin (1975) discussed it as a parasitic strategy. Examples from some groups are given in the following. The ciliate *Orchitophyra stellarum* parasitizes the testes of the starfish *Asterias rubens* near Plymouth, England, and completely destroys the germinal tissue of the testes with resulting castration (Vevers 1951); the same parasite also infects the ovaries of a related starfish, *Asterias vulgaris*, without any apparent damage to the egg cells (Smith 1936). Chaetognatha infected with trematode larvae in the Mediterranean were found to be partially castrated (Pearre 1976). Parasitic castration is particularly common in the molluscs. All known sporocysts cause it in the marine mussel, *Donax* (see Hopkins 1958), and parasitized snails are usually sterile (Ewers and Rose 1965). Rohde and Sandland (1973) found that most snails of the species *Cerithium moniliferum* at Heron Island, Great Barrier Reef, infected with trematode larvae were sterile. Of 135 snails without trematodes, 10 (15 per cent) had eggs, of 81 snails with trematodes, only 1 (1 per cent) did. None of the 41 snails with both trematodes and the aspidogastrid *Lobatostoma manteri* had eggs. Barnacles of the genus *Sacculina* commonly sterilize crabs (Haswell 1888, Phillips and Cannon 1978).

Occasionally invertebrate hosts infected with parasites are abnormally large, a phenomenon known as gigantism. Pearre (1976), for instance, found that Chaetognatha in the Mediterranean infected with trematodes showed gigantism, and the phenomenon was demonstrated in snails of several species at Plymouth infected with trematode larvae (Rothschild 1936, 1938, 1941a,b). An increase in growth rate during the early stages of infection has been suggested for oysters infected with the trematode, *Bucephalus cuculus* (Menzel and Hopkins 1955) and for the snail, *Littorina neritoides*, infected with trematodes (Rothschild 1941a).

Some hosts are remarkably little affected. Davis and Farley (1973), for instance, found no significant difference in the efficiency of carbon and nitrogen assimilation and in ingestion in the snail, *Littorina saxatilis*, infected with rediae of *Cryptocotyle lingua* compared with uninfected ones. The digestive efficiency in the related snail, *L. littorea*, was also not changed by infection with the same parasite

(Platt 1968). However, the apparent lack of effect by the parasites may simply be due to the methods not being sensitive enough to detect such effects, or effects may be detectable only under certain conditions.

Effects on vertebrates More numerous than studies dealing with pathological effects on invertebrate hosts are those concerned with effects on vertebrates, particularly fish. Details can be found in Schäperclaus (1954), Kabata (1970), Sindermann (1966, 1970), Reichenbach-Klinke and Landolt (1973), and Needham and Wootten (1978). A bibliography of parasites of marine and freshwater fishes of India and the diseases caused by them was given by Natarajan and James (1977). Reichenbach-Klinke (1973) reviewed the effects of parasites on fish hosts. Although his examples are mainly tapeworms of freshwater fishes, it seems probable that similar effects occur in marine fish, which are less well known. The main effects, as discussed by Reichenbach-Klinke, are as follows.

Intestinal parasites inhibit the digestive activity of the hosts and indirectly inhibit vitamin and blood sugar metabolism. Parasites in the liver affect glycogen metabolism and growth, whereas parasites of the gonads and coelomic cavity may lead to complete castration. Cestode larvae (*Triaenophorus*) in the liver may lead to retardation of growth, and cestode larvae in the muscles may cause a loss in weight of 10 per cent or more. Reduction in egg numbers have so far been found to be due only to parasites of the body cavity.

Only a few examples will be discussed here, some from freshwater fish because they are better known than marine species. The severity of pathological effects varies greatly, from nil or almost nil to death. Monogenea feeding on the epidermis of fish, for example, rarely cause distinct symptoms, apparently because the epidermis has a rapid rate of regeneration and fast sloughing of epidermal cells prevents any lasting damage (Kearn 1963b). But even damage due to Monogenea varies considerably. According to Bychowsky (1957), Monogenea with clamps apparently inflict the least damage. Various species of parasite affect different fish species differently. Ecological factors, specific behaviour patterns, age of the host and specific characteristics of Monogenea determine the rate and intensity of infection and with it the damage caused. Sometimes the damage may be indirect, as suggested by Remley (1936, 1942) for *Aplodinotus grunniens* infected with *Microcotyle spinicirrus*. The parasite does not cause extensive damage to the gills, but probably breaks down the mucous layer and enables fungus to invade and kill the host.

Lester and Adams (1974) gave experimental evidence for a relation between number of the monogenean *Gyrodactylus alexanderi* and its host fish, *Gasterosteus aculeatus*. Heavily infected fish frequently died. Over one to two weeks they became sluggish, emaciated and finally moribund. Of 174 infected fish with less than 150 worms per fish, only 33 died, whereas 54 of 67 fish died after infection reached 150 worms, usually within the next few weeks when the number of flukes had risen to 300–400. The maximum number on one moribund fish was 700 worms. But except for hyperplasia of the epidermis, no signs of injury were noticed. Pathological conditions associated with other species of the monogenean family Gyrodactylidae are increased mucus production, shrivelled gills, loosened scales, haemorrhages, swelling skin and frayed fins (Lester and Adams 1974).

Pathological conditions associated with several parasitic species were observed by Paperna and Zwerner (1976a,b) in the estuarine fish, *Morone saxatilis*, from the lower Chesapeake Bay, U.S.A. According to these authors, the copepod, *Ergasilus labracis*, is responsible for extensive hyperplasia of the gills, the nematode, *Philometra rubra*, causes granuloma of the viscera and adhesions, and infection with the acanthocephalan *Pomphorhynchus rocci* leads to tissue reactions in the intestinal wall. Furthermore, various larval helminths cause an extreme fibrosis especially of the liver and spleen, and the crustacean, *Argulus bicolor*, causes skin lesions. The authors did not consider it likely that the infections cause mass mortalities, but suggested that heavily infected fish may die, particularly when the water quality deteriorates. They also suggested that worm infections may contribute to juvenile fish mortality and that reduced infection rates of old fish are due to selective mortality. Infections of the winter flounder, *Pseudopleuronectes americanus*, with a microsporan strongly suggest that fishes heavily infected during the first year of life do not survive into their second year (Stunkard and Lux 1965). The great range of pathological effects is indicated by McVicar's (1972) observation that two tapeworms, *Echeneibothrium* sp. and *Phyllobothrium piriei*, severely damage the intestinal mucosa of their host, whereas a third, *Acanthobothrium quadripartitum*, does little harm. The differences are probably associated with the size, motility and mode of attachment of the parasites. The copepod, *Elytrophora brachyptera*, causes holes in the skin near the pleurobranch and formation of large blood lacunae in the tuna, *Thynnus thynnus* (Lüling 1953), but many copepods, for example various species of *Caligus*, cause little or no damage (personal observations).

Effects on host populations; mass mortalities

There are several recent reviews on factors which regulate parasite populations and determine the outbreak of epidemics (epizootics). Kennedy (1975) included a chapter on "epidemiology and models of host–parasite systems" in his book *Ecological Animal Parasitology*, and the same author (1977) discussed regulation of fish parasite populations. Whitfield (1979) devoted a brief chapter of his book to the regulation of parasite and host populations, and Anderson and May (1979) and May and Anderson (1979) used a mathematical approach in their discussion of the population biology of infectious diseases, including protozoan and metazoan parasites.

In general, authors with a theoretical background tend to emphasize the importance of "regulatory mechanisms of the density dependent type" (Anderson 1976). Such regulatory mechanisms are supposed to be intraspecific competition of larval trematodes in mollusc hosts, constraints on adult flukes by the host's immunity response and mortality of hosts caused by parasites, all of which are said to stabilize the host–parasite relationship (Anderson 1976). Models based on theoretical assumptions also permit predictions on how host populations should react under certain conditions. Thus Anderson (1980), using simple population models to study the effect of parasites on populations, predicted that parasites of low to intermediate pathogenicity suppress host population growth most effectively, whereas highly pathogenic species more probably cause their own extinction but not that of their host. Furthermore, host populations with high reproductive potential can withstand the effect of parasites better, and parasites which multiply within the host suppress population growth of the host most strongly and cause their own and their host's extinction. Host density, rates of host and parasite reproduction, the statistical distribution of parasite numbers per host, and density-dependent constraints on parasite population growth are also important.

On the other hand, authors more impressed by empirical data tend to emphasize "density-independent" factors and the instability of host–parasite systems (Kennedy 1975, Price 1980). According to Kennedy (1975), "the most striking feature of intermediate host–parasite systems is . . . instability"; incidence and intensity are almost always determined by climatic or ecological factors influencing the probability of contact. With regard to vertebrates, "with the conspicuous exception of mammals, which possess well-developed immune systems, mortality is seldom due to specific responses by the host against the parasite. Instead, low probability of infection, variation in susceptibility of hosts to infection and non-specific mortality during invasion prevent the build up of heavy infections."

Few data are available on infection intensities and frequencies in "normal" populations of marine organisms, but it is clear that parasite numbers under natural conditions are usually far below that which a host individual and population could carry, as evidenced by the heavy parasite loads occasionally observed.

Ectoparasites of marine fish are usually rather rare (Rohde 1979b, c, 1980d, Rohde *et al*. 1980). The same applies to their endoparasites as well, as indicated by numerous dissections (Rohde, unpublished). These findings imply that "density-dependent" factors, factors resulting from the numbers of individuals in a population, do not play a role or only a very minor one in regulating host populations. Even if saturation of a host individual with a parasite species is reached, this may be due to limits imposed by the parasite's microhabitat and represent no danger to the host. For example, the cestodarian *Gyrocotyle* in chimaerid fishes occurs usually in infection intensities of two adults and the same was shown for the trematode, *Deretrema*, in the teleost fish, *Anomalops katopteron* (Burn 1980). The worms are large compared with their microhabitat in the fish, and more than two worms apparently cannot be accommodated.

The previous discussion has shown that effects of parasites on host individuals are sometimes deleterious. But even if individuals are castrated or die, host populations are not necessarily greatly affected. James (1965) examined the effect of five species of cercariae on a population of the snail *Littorina saxatilis* with two reproductive periods. Even though infections led to the destruction of the gonads and sometimes to death, some species of cercariae had little or no long term effect on the host population, as snails were affected only after the first or second reproductive period. Furthermore, even reduction of offspring by an infection may not cause a host population to decline, if enough offspring are left to replace adult losses.

Corresponding to the low infection rates of hosts with most parasites, observations of mass mortalities under natural conditions are rare. Even in cases where conditions seem to be normal, the effects of human activities, such as overfishing or removal of predators, cannot be ruled out. For example, mass mortalities of mullet in the Azov and Black Seas due to parasite infections (see page 157) may well be the consequence of such activities. On the other hand, mass mortalities under abnormal, crowded conditions, have been observed frequently. Paperna (1975) reported severe injuries and mortalities associated with heavy infections of the monogenean *Benedenia* in mullet kept in tanks, and Paperna (1978) observed epizootics in several cultured fish species due to the protozoans *Trichodina* and *Amyloodinium ocellatum*, and the monogeneans *Furnestia echeneis*, *Bivagina* sp. and *Benedenia monticelli*. The ascetoporan *Minchinia nelsoni* caused an epizootic in oysters at the mouth of the York river, Virginia, in 1959, leading to heavy losses and abolishment of commercial oyster planting by 1960 (Andrews and Wood 1967). Another ascetoporan parasite of oysters, *Minchinia costalis*, caused short, sharp outbreaks in May–June of three consecutive years (Andrews *et al.* 1962, Wood and Andrews 1962). Other examples can be found in the chapter on disease of marine animals (see also Sindermann 1963 and Sindermann and Rosenfield 1967).

One important factor responsible for the outbreak of epizootics is an abnormally high density of the host population. For example, Korringa (1952a) found that the copepod, *Mytilicola intestinalis*, occurs in dangerous numbers and seriously interferes with health, growth and fattening of its mussel hosts only in those districts where mussel populations are dense. Vevers (1951) observed that the ciliate, *Orchitophrya stellarum*, infects only "numerically rich and well-fed" populations of the starfish, *Asterias rubens*, near Plymouth, England, but never geographically adjacent populations with smaller numbers of poorly fed starfish.

Density of host populations may also affect the number of parasitic species. According to Fenchel (1966), bivalves which usually have the densest populations harbour more species of ciliates in their mantle cavity than rare species. Thus, the common mussels, *Mytilus edulis* and *Macoma balthica*, have 7 and 6 species of ciliates respectively, whereas rare or sporadically occurring host species have either few species or none.

On the other hand, several authors have shown that the efficiency with which a parasite locates and infects hosts, may decrease with increasing density of host and parasite (for references see Chua 1979). This indicates that there may be no simple linear relationship between population densities and infection rates, and that different species may behave differently in this respect. No data are available from marine parasites, but observations of the much better known parasites of man and domestic animals show that the exact relationship between infection success and population densities may well depend, at least in part, on how the infective stages locate the host, whether there is direct transfer from parasite to parasite, and whether intermediate hosts are involved.

Beside population density, many other factors may determine infection rates, spreading and mortalities, and these factors may vary from host to host. With regard to *Mytilicola intestinalis*, a copepod of mussels for example, Korringa (1968) pointed out that the number of parasites per host individual is mainly determined by the number of hosts living in a given volume of water, but is also influenced by the amount of flushing of the water. According to Bolster (1954), factors responsible for the degree of infection by this species are salinity, depth of water, turbulence and speed of current. Hepper (1955) stated that infections are heavier in mussels from the sea bottom, slow moving water and the middle region of estuaries than in mussels raised from the sea bed, in fast moving water and at either end of an estuary.

Spreading into new areas by this parasite mainly occurs as a result of transport of infected mussels on the bottom of ships (Bolster 1954). This agrees with the observation of Grainger (1951) that all four of 21 samples of the mussel, *Mytilus edulis*, from various parts of Ireland, infected with the parasite, were from the neighbourhood of ports.

Even in host populations under abnormal conditions, it is often doubtful whether mass mortalities are solely or even mainly due to a parasite. For example, Laird (1961) doubts that parasites as such are responsible for large scale oyster mortalities. He states that "no studies of major oyster mortalities have yet proved any specific parasite of first importance *per se*, and it has been contended that such parasites only assume grave significance when general environmental deterioration so lowers host resistance as to transform regular low-level loss into disaster." He points out that oysters are often grown "under conditions inviting intermittent disaster". Although his data are admittedly scanty and mainly qualitative, he believes that they support the conclusion that disease caused by the flagellate, *Hexamita* (and by other pathogens, see Shuster and Hillman 1963), is caused when the normally saprozoic organism becomes a facultative parasite, when temperature, rainfall or tidal factors render the local environment unfavourable to the oysters and lower their resistance to infection. Thus, oyster hexamitiasis is "simply an end point in a chain of events".

Farley (1968) arrived at similar conclusions concerning the oyster parasite, *Minchinia nelsoni*. Mortality may be partly attributed to environmental or physiological stress, which infected and therefore weakened oysters cannot tolerate. High salinity appears to favour outbreak of the disease, indicated by the observation that from 1963 to 1966 during a drought period in Virginia, U.S.A., mortalities were higher (Andrews 1968). Densities of oyster populations did not appear to be important in this case, as oysters were extremely scarce and mortalities nevertheless high (Andrews 1968).

Pearse and Wharton (1938) also noted that the oyster "leech", *Stylochus inimicus*, a predator rather than a parasite, may kill oysters, but is probably rarely, if ever, the sole cause of mass mortalities. Similarly, Korringa (1957), analysing the decline in mussel production in parts of the Netherlands in 1954–1956 (see page 167), held that the great average age and the poor quality of the mussels planted in early 1955 were mainly responsible although the copepod *Mytilicola* also appeared to be a causative agent of extensive mussel mortalities (Korringa 1968).

Sindermann and Rosenfield (1967), in their review on disease of marine fish and shellfish, concluded that many physical, chemical and biological variables contribute directly or indirectly to mortality. Often, the cause is undetermined and disease is only suspected. The contention that disease ranks first among all causes of mortality of oysters, made by Machin (1961, cited by Sindermann and Rosenfield 1967) remains to be substantiated. Based on the largely rudimentary information available, arising almost entirely from the study of commercially important species or cultured species, both usually occurring in high densities, Sindermann (1963) listed the following factors influencing epizootics of marine animals: population explosion of an introduced pathogen to which a host is susceptible; changes in the physical environment of the host population; changes in virulence and infectivity of a pathogen already present; changes in the effectiveness in the transmission of a pathogen enhanced by high population densities; changes in the susceptibility of the host population. He drew attention to curious synchronizations of outbreaks of several diseases in different marine species.

In spite of the fact that often more than just the presence of a pathogen is needed for the outbreak of a disease, the importance of parasites should not be underestimated. There can be no doubt that parasites such as *Minchinia nelsoni* and *Marteilia refrigens* have damaged oyster farming in several regions, and it seems wiser to overemphasize the importance of such agents than to underestimate it, at least until all factors responsible for epizootics have been closely studied.

Co-evolution of hosts and parasites; use of parasites for the study of host phylogeny and origin

Eichler (1948), partly on the basis of earlier observations of certain regularities in the distribution of parasites in various groups of hosts, formulated three rules governing evolution of parasites with respect to their hosts. They were Fahrenholz's rule, Szidat's rule and the divergence rule (later called Eichler's rule by others).

1. Fahrenholz's rule states that the classification of some groups of parasites parallels that of their hosts, which implies that the ancestors of extant parasites must have been parasites of the ancestors of extant hosts, or in other words that the evolution of host and parasites must have been parallel.
2. Szidat's rule states that "primitive" hosts are parasitized by "primitive" parasites and specialized hosts by specialized parasites.
3. According to Eichler's rule, large host groups have more genera of parasites than small groups, or "isolated groups of hosts often do not harbour many kinds of parasites, whilst in comparable groups of hosts with many species or near affinities to other groups of hosts we may find not only many species . . . but also differentiation into many genera which may live together on the same host species".

All the evidence given by Eichler is from non-marine groups, and relatively little evidence from marine parasites can be offered because they are less thoroughly known.

The rules are not equally valid for all groups of parasites. Generally, the lower the host specificity, and particularly the phylogenetic specificity (see Chapter 7), the smaller the applicability of the rules. The reasons are obvious: non-specific parasites infect hosts irrespective of their phylogenetic status.

The often close correspondence of host and parasite evolution (Fahrenholz's rule) permits the use of parasites in the study of host phylogeny (Eichler 1948). The seemingly "retarded" evolution of parasites, due to the fact that the simpler structure of parasites often prevents evolutionary changes as great as those of their hosts, may result in a greater similarity of related parasites than of equally related hosts and, thus, facilitate tracing of host phylogeny. Bychowsky (1933), for instance, used monogeneans for the study of cyprinid taxonomy. On the other hand, parasites may diverge more clearly than their hosts and thus permit differentiation of closely related host species (Mayr *et al.* 1953, cited by Price 1980) or host populations. Red Sea and Australian populations of the ray *Taeniura lymna*, for instance, have two different species of the cestode *Anthobothrium* and some other cestodes are also different. Williams (1964), on this basis, suggested that fish from intermediate regions should be examined to clarify whether different subspecies of *Taeniura* exist.

Terrestrial and freshwater parasites and ectocommensals have been used in attempts to clarify not only relationships of hosts, but also places of origin and dispersal of hosts, and ancient land connections between land masses (Metcalf 1929). The method was first used by von Ihering (1891, 1902) and is therefore often referred to as the "von Ihering method". One assumption on which this method is based is that groups of animals usually have a greater diversity in areas where they have been for a long time, normally where they have originated. With regard to parasites, the reasoning is that hosts are likely to have acquired the greatest variety of parasites in the region where they have lived the longest, and that parasites in such regions had, in turn, much time to become more and more specialized and host specific. If a host moves into a new area, it will lose many of its parasites, especially those with indirect life cycles (Manter 1967). Only few applications of the method to marine animals have been made.

The study by Szidat (1961) of the parasites of the hake, *Merluccius hubbsi*, which is restricted to the Argentine coast, is an example. It had been thought to have evolved from *M. merluccius* or *M. bilinearis* in the north Atlantic, but all its parasites are similar to those of the north Pacific hake, *M. productus*, and unlike those in the Atlantic. Szidat concluded that *M. hubbsi* is of Pacific origin. He termed parasites which indicate some ancient condition "Leitparasiten" (translated as "index parasites" by Manter 1966), corresponding to the index fossils of paleontologists. Manter (1966) made a study of parasites of the marine fish genus *Kyphosus* from various zoogeographical regions. Six of the 11 species of *Kyphosus* occur in the Indo-Pacific, and only 2 in the Caribbean and the same 2 in West Africa. Twenty-one species of trematodes are known, and most trematode genera are limited to *Kyphosus*. Manter found that 3 or 4 species of trematodes of *Kyphosus* from the Great Barrier Reef in northeastern Australia and from the Caribbean are identical, a similarity greater than that between any other two regions. Furthermore, 1 acanthocephalan is more or less identical with a species in the Gulf of Mexico, and 1 monogenean is also similar to a species in that region. On the other hand, the species *Kyphosus sydneyanus* extending from southern Australia almost to the Great Barrier Reef has 4 trematode genera and species of which only one is like those from the Reef. Manter (1967) interprets this as meaning that (1) hosts and parasites have originated in the Indo-Pacific (because of the larger number of species there); (2) there was an early dispersal to South Australia, where isolation has been rather complete for a long time (indicated by the different parasite fauna and also by the remarkably distinct trematode fauna of marine fish in general); and (3) there has been some broad main dispersal road to the Caribbean.

Another example of the use of parasites for the study of host origin is Manter's (1955) study of parasites of eels from the Atlantic and the Pacific. He found that 20 species of trematodes of Atlantic eels comprise 14 species which are shared with other marine fishes, that is are non-host specific, 5 which are shared with freshwater fishes, and only one species which occurs in eels only. Eels from the Pacific are known to have 3 genera and 8 species of trematodes which are specific to them. He suggests that the large number of specific parasites in the Pacific indicates a Pacific origin of eels (*Anguilla*), whereas the large number of parasites shared with marine fishes in the Atlantic indicates that eels there had longer contact with the sea. However, his conclusions may have to be modified once more studies in the Pacific have been made. Parasites of eels in the Atlantic are better known than those of eels in the Pacific.

CHAPTER 7

The Ecological Niches of Parasites

The niche concept

There is no general agreement on what a "niche" is, but most authors define a niche as the total of an organism's relations to its environment, biotic and abiotic (for a recent discussion see Pianka 1976). These relations determine the organism's place in nature, or in ecological jargon, its place in "multidimensional hyperspace". For example, the niche of an animal species is defined by certain requirements for space, time, food, temperature, oxygen concentration and other factors, and these requirements can be looked at in terms of gradients along various niche dimensions. Although their number seems to be almost infinite, there is much overlap between the various dimensions and a few are sufficient to describe an organism's niche with a great degree of accuracy.

The niche of an organism may be influenced by the absence or presence of competing species, that is species with overlapping requirements. The "fundamental niche", in the absence of competitors, may be compressed to a "realized niche" if competitors are present. Removal experiments can demonstrate the effect of competition. For instance, competition must be responsible for narrow microhabitats, if one of two copepod species living on the same fish individual expands into the microhabitat of the second species after the latter's removal. Its niche must have been restricted previously by the other species. The same effect can be shown if fish individuals are compared which are naturally infected with only one or with both species.

Niche dimensions of parasites

As already pointed out, the number of aspects of a niche (its dimensions) are almost infinite, and it is not practicable to discuss them all. However, a few are usually sufficient to characterize a species' niche. Among the best known marine parasites are the Monogenea, and Rohde (1979c) discussed the most important aspects of their niches. They are host specificity, geographical range, macrohabitats, microhabitats, sex of host, host age, season and hyperparasitism.

Host specificity Host specificity is the restriction of parasites to certain host species. It is universal, although its degree differs among different parasites. Host specificity in various animal groups was discussed by Manter (1957, Digenea of marine fishes), Llewellyn (1957, Monogenea), Golvan (1957, Acanthocephala), Dollfus (1957, tetrarhynchid cestodes) and Euzet (1957, cestodes of sharks). Table 4 contains some examples which show that Monogenea have the highest degree of host specificity among parasites of marine fishes. Eleven of the 21 species in the table are limited to a single host species, and only one sometimes occurs on hosts of different families. The high degree of host specificity of marine Monogenea is also indicated by the data given by Rohde (1978f, 1979c). Of 435 species of Monogenea from various seas, 340 (78 per cent) are restricted to one host species, 388 (89 per cent) to one genus, 420 (96 per cent) to one family, and 429 (98 per cent) to one order. Monogenea have even been used to differentiate host species (Mizelle et al. 1943). All other groups of which a sufficient number of species have been examined, are less specific. Of the examples in Table 4, 5 of 10 species of nematodes, 7 of 17 species of protozoans, and 22 of 36 species of cestodes in which specificity was determined, parasitize more than one family of hosts.

Host specificity of copepods of marine fishes was examined by Boxshall (1974). Among 39 species of copepods on 41 species of fish, 70 per cent were specific for one host species or for several species in one genus. Of 42 species of ciliate protozoans living in the mantle cavity of bivalves, 21 were strictly host-specific (Fenchel 1966).

Parasites which infect a single host taxon or related taxa, are said to exhibit a phylogenetic host specificity. Even parasites with a wide host range always show certain host preferences, but these preferences are usually determined by the ecological requirements of the host. Such parasites are said to exhibit an ecological host specificity. Groups consisting of members with strict phylogenetic host specificity may also contain some species which have secondarily widened their host range, and have acquired an ecological host specificity. Examples may be found in the monograph on Monogenea of Hawaiian fishes by Yamaguti (1968). According to him, 71 per cent of monogenean species found during the examination of 2,097

Table 4. Host specificity of parasites of fish in the Barents Sea. Host records from other seas are considered, records of accidental hosts in which parasites do not mature are not. Data from Polyanski (1966).

Parasite group	Number of species	In 1 host species	In > 1 species of 1 genus	Percentage of species: In 1 family	Primarily in 1 family	In several families	Undetermined
Protozoa	25	21.7	4.3	17.4	21.7	8.7	26.2
Monogenea	21	52.4	9.5	33.3	4.8	0	0
Digenea	37	2.8	11.1	25.0	16.7	44.4	2.8
Cestoda	19	12.5	6.2	18.7	25.0	31.4	6.2
Nematoda	12	9.1	0	36.3	9.1	36.4	9.1
Acanthocephala	3	0	0	0	0	100.0	0
Hirudinea	3	33.3	33.3	33.3	0	0	0
Copepoda	15	6.7	20.0	27.0	33.2	6.7	6.7
Isopoda	1	0	0	0	100.0	0	0
Total	136	17.2%	9.2%	24.9%	17.9%	23.9%	6.9%

individuals of 122 fish species were restricted to a single host species, 85 per cent to one or two host species, and only 11 (8 per cent) occurred in more than one family of host. Several species of the genus *Benedenia* were limited to one host, but the species *B. hawaiiensis* was recorded on 24 host species. Most of these fish belong to different families, but they live in similar habitats (coral reefs). Another example from the same monograph is *Enoplocotyle hawaiiensis*, which was found on eight species of reef fish.

In many cases, our knowledge of the host range of parasites is insufficient, and parasites originally thought to exhibit a strict host specificity are later found to be less specific. Thus, some early authors claimed a great host specificity for most species of the snail family Pyramidellidae, but it now seems that most of them are probably non-specific, attacking a range of hosts (Ankel and Christensen 1963).

On the other hand, even parasites found on or in many host species usually infect one or a few species more heavily than others. Thus, the copepod *Mytilicola porrecta*, although infecting several species of bivalves, shows clear preferences for certain host species (Hepper 1956), and experiments showed that neither of the pyramidellid snail species, *Odostomia seminuda* and *O. bisuturalis*, has a strict host specificity, although each species has a certain order of preference for certain hosts (Boss and Merrill 1965).

To account for frequency and intensity of infection in various hosts in addition to the host range, Rohde (1980e) developed host specificity indices. One such index uses the number of parasite individuals found in each host species (the "parasite density"). He defined the host specificity index based on parasite density as

$$S_i \text{ (density)} = \frac{\sum \frac{x_{ij}}{n_j h_{ij}}}{\sum \frac{x_{ij}}{n_j}}$$

where S_i = host specificity of the ith parasite species, x_{ij} = number of parasite individuals of ith species in jth host species, n_j = number of host individuals of jth species examined, h_{ij} = rank of host species j based on density of infection $\frac{x_{ij}}{n_j}$ (species with greatest density has rank 1).

If all parasite species of a community are considered, the specificity index of the parasite community can be defined as

S_c (density) $= \dfrac{\sum S_i}{n_p}$ where n_p = number of parasite species in the community.

Using the same formula, a host specificity index based on frequencies of infection can be defined. Parameters in the formula are now changed as follows. x_{ij} = number of host individuals of jth species infected with parasite species i, n_j = number of host individuals of jth species infected, h_{ij} = rank of host species based on frequency of infection. One can also make use of densities as well as frequencies of infection, for instance by using a combined index

$$\dfrac{S_c(\text{density}) + S_c(\text{frequency})}{2}.$$

Numerical values for the indices vary between 0 and 1. The closer to 1, the higher the degree of host specificity, the closer to 0, the wider the host range. Application of the indices to Digenea from the White Sea shows that, in spite of the wide host range of many species, most infect only a few host species heavily. *Lecithaster gibbosus* was found in 12 of 31 fish species examined in the White Sea (Shulman and Shulman-Albova 1953); its S_i (frequency) is 0.54 and its S_i (density) 0.99 (Rohde 1980e). The large S_i (density) indicates that the vast majority of all parasites of this species is found in a single host species. Further examples are demonstrated in Fig. 60.

A different index of host specificity was proposed by Price (1980). To characterize the host specificity of a parasite group, he defined as "per cent specificity" the percentage of species in a family that utilize only one host.

It may be useful to distinguish host specificity from host range. Host range is the number of host species found to be infected by a certain parasitic species irrespective of how heavily and frequently the various host species are infected. Host specificity takes intensity and/or frequency of infection into account.

Microhabitats As universal as the preference for certain hosts is the preference for certain microhabitats on or in the hosts. Ecto- and endoparasites of invertebrates and vertebrates all show such preferences. The sporulating stages of the ascetosporan *Minchinia nelsoni* are restricted to the digestive diverticula of oysters (Couch 1967). The aspidogastrean trematodes, *Aspidogaster* sp. and *Lophotaspis margaritiferae*, live in the pericardium of marine bivalves (Shipley and Hornell 1904, Bychowsky and Bychowsky 1934, Moriya 1944), and *Lobatostoma manteri* nearly always lives in a certain part of the digestive gland of the marine snail *Cerithium moniliferum*, and in the stomach and ducts of the digestive gland of *Peristernia*

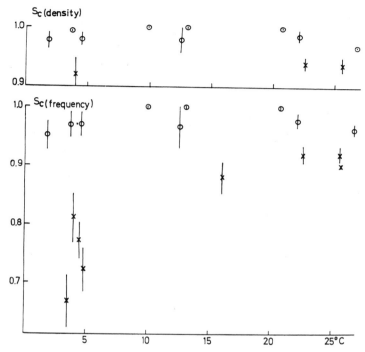

Fig. 60. Host specificity indices of marine Monogenea (o) and trematodes (x) at different latitudes as characterized by approximate means of annual sea surface temperature ranges (after Rohde 1980b).

australiensis (Rohde 1973, Rohde and Sandland 1973). Parasites in the digestive tract of marine fishes occupy more or less restricted microhabitats, and they are never randomly distributed along the whole tract (Williams 1968b). Some examples are shown in Fig. 61. The same refers to endoparasites of other tissues and organs, and to ectoparasites on the skin and gills. For example, the blood fluke, *Aporocotyle macfarlani*, lives in the lumen of the afferent branchial arteries, the vertral aorta and occasionally parts of the heart of scorpaenid rockfishes, *Sebastes* spp. (Scorpaenidae), on the North American Pacific coast, and *Psettarium sebastodorum* parasitizes the intertrabecular spaces of the ventricle and atrium of the same species in the same area (Holmes 1971). Shotter (1976) found that all helminth and copepod parasites of the whiting, *Merlangius mer-*

langus, showed preferences for certain tissues or parts of the digestive tract. MacKenzie and Gibson (1970) discussed site specificity of helminths in various organs and tissues of 500 plaice, *Pleuronectes platessa*, and 900 flounder, *Platichthys flesus*.

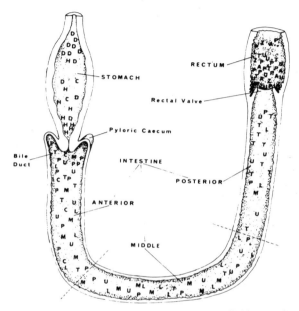

Fig. 61. The flounder gut opened to show the regional divisions and parasite distributions: A, *Pomphorhynchus* sp.; C, *Thynnascaris aduncum*; D, *Derogenes varicus*; H, *Hemiurus communis*, L, *Lecithaster gibbosus*; M, *Cucullanus minutus*; P, *Podocotyle* sp.; T, *Tetraphyllidean* larvae; U, *Cucullanus heterochrous*; Z, *Zoogonoides viviparus* (after MacKenzie and Gibson 1970).

Microhabitats of monogeneans have been studied particularly well (Stunkard 1922, Achmerov 1954, Llewellyn 1956, Bogdanova 1957, Bychowsky 1957, Robinson 1961, Pratt and Herrman 1962, Young 1968, Gusev and Fernando 1973, Rohde 1976b,c, 1977a,b,c, 1978b, 1979c, 1980d, Roubal 1979, Armitage 1980, Byrnes 1980). Figure 62 illustrates the distribution of two species of Monogenea as well as one species of copepod and cysts on a species of mackerel.

Rohde (1977c) showed that ectoparasites on the gill of marine fish may partition their microhabitats in the following ways. There may be: (1) transverse partitioning (preference for certain gill arches); (2)

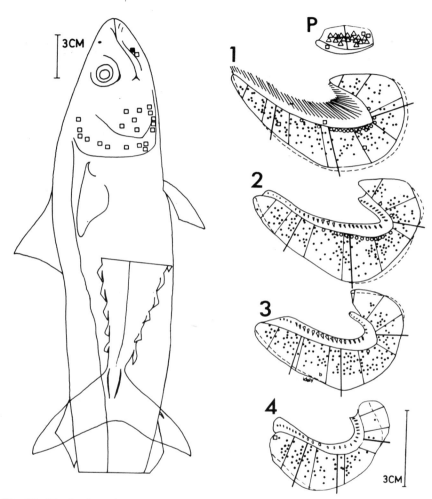

Fig. 62. Distribution of ectoparasites on the surface and in the mouth cavity of 122 *Scomber scombrus* at Helgoland, North Sea.
  ◌ *Caligus pelamydis* in mouth cavity and on gills,
  ■ *C. pelamydis* in external fold of mouth;
  ● cysts;
  ○ *Kuhnia scombri* (1 circle = approx. 5 individuals);
  ∆ *Kuhnia* sp. (after Rohde 1980d).
  P = pleurobranch, 1-4 = gills nos 1-4.

longitudinal partitioning (preference for certain microhabitats along the longitudinal axis of the gills); (3) vertical partitioning (preference for certain microhabitats along the axis extending from the tip of the gill filaments to the bony part of the gills); (4) lateral partitioning (preference for external or internal gill filaments). The parasites also may prefer the anterior or posterior surface of the gill filaments, and it even seems possible that some gill parasites show a preference for the left or the right side of the body. This is not surprising in view of the asymmetrical body shape of many gill parasites, particularly Monogenea. Preference for one side of the body cannot be expected in those species which can adapt their asymmetry easily to the side of the body where they arrive. But it can be expected for those species which have a genetically rigid asymmetry and consist of populations with entirely or predominantly one type of asymmetry. In monogeneans, asymmetry is sometimes induced (Ramalingam 1960), but may sometimes be inherited (Young 1968, Lebedev 1969a). A distinct preference for left or right, however, so far has not been demonstrated beyond doubt in monogeneans, although it has been claimed for *Diplozoon paradoxum* on freshwater fish by Wiles (1968). With regard to copepods, Larraneta (1957) stated that the species *Peroderma cylindricum* distinctly prefers the left side of sardines off Spain (cited by Mann 1970).

That microhabitats are not always static is shown by the example of the copepod *Clavella uncinata* on the whiting, *Merlangius merlangus*. In young fish, over two thirds of the copepods were attached to the posterior rim of the operculum, whereas in older fish most copepods were attached to the gill rakers (Shotter 1976). Frequently, larvae and adults have different microhabitats, for instance in the copepod, *Caligus diaphanus*, on *Trigla lucerna* in the North Sea. Larvae and young females and males live on the gill filaments, whereas adults inhabit the inner wall of the opercular cavity (Rohde 1980d).

Macrohabitats The macrohabitat of a parasite includes those niche components which also represent the habitat of its host(s). A sandy beach and an adjacent rocky shore, or the estuary of a river and adjacent deeper sublittoral represent different macrohabitats. Sometimes, a parasite species may occur only in some of the macrohabitats inhabited by its host, sometimes it may infect several host species with different macrohabitats, that is its macrohabitat range may be narrower or wider than that of its host(s).

Examples of species showing macrohabitat preferences are the nematode larvae of marine fish in southeastern Queensland, Australia, examined by Cannon (1977a,b). He found nine distinct larval types of four ascaridoid genera in 47 of 123 fish species. One genus, *Anisakis*, is restricted to open water fish, another, *Contracaecum*, to inshore shallow water fish, and the remaining two, *Phocanema* and *Thynnascaris*, have intermediate distributions. Definite macrohabitat preferences were also shown for some parasites of the sole, *Parophrys vetulus*, in Oregon, U.S.A. The microsporan *Glugea stephani*, the acanthocephalan *Echinorhynchus lageniformis*, and the nematode *Philometra americana* were acquired only in the estuary, and the trematodes *Otodistomum veliporum* and *Zoogonus dextrocirrus*, the leech *Oceanobdella* sp. and three species of copepods were acquired only offshore (Olsen 1978). Different species of boring sponges, *Cliona* spp., on oysters, show different preferences for certain salinity zones. As a result of a long drought in South Carolina, U.S.A., high salinity species increased in places where low salinity species had been more abundant previously (Hopkins 1956a). Another parasite of oysters, the polychaete, *Polydora ciliata*, was present in about 50 per cent of oysters above the low water mark, and in about 86 per cent of oysters below the low water mark in South Carolina. Although data were insufficient to separate the effect of differences in salinity and depth in this case, generally *Polydora* appears to be more prevalent in waters of low salinity. A marked effect of the substratum on the degree of infection was apparent. On firm or hard bottom above low water mark about 21 per cent of the oysters were infected, on muddy or soft bottom at the same level about 52 per cent were infected (Lunz 1941).

Geographical range Like the macrohabitat dimension, the geographical range of a species is a dimension in space and it is not always possible to distinguish both clearly. Parasites may have a wider geographical range than any particular host species, if they infect different hosts in different areas, or they may have a narrower geographical range, if one or several host species are infected only in part of their range. An example of a species having different hosts in different areas is the monogenean *Octodactylus minor*. In the Barents Sea, it infects the fish *Micromesistus poutassou* only, but it also occurs on the Atlantic cod, *Gadus merlangus*, on the Norwegian and Irish coasts. Similarly, the monogenean, *Diclidophora denticulata*, infects only *Pollachius virens* in the Barents Sea, but also *Merluccius merluccius* and *Gadus minutus* in the European and American Atlantic (Polyanski 1966). On the other hand, the monogenean *Pseudothoracocotyla gigantica* has been found so far only near Heron Island, Great Barrier Reef (Rohde 1976b), although its host, the Spanish Mackerel, *Scomberomorus commerson*, is found in large parts of the Indian Ocean. However, as in most cases of such restricted records, it is possible that it will be found in other areas in the future.

Sex of host Only in a few cases has a preference for one sex of the host by monogeneans been demonstrated. Paling (1965) found that a particular population of the monogenean *Discocotyle sagittata* on *Salmo trutta* affects 5–7 year old males more heavily than females, and according to Williams (1965), *Calicotyle kroyeri* is never present in gravid female rays, *Raja radiata*, although non-gravid females may be infected. Males of the snail, *Hydrobia ulvae*, in Britain are more commonly infected with larval trematodes than are females, the ratio of infected males to infected females sometimes reaching 16:1 (Rothschild 1938). Metacercariae of another trematode also are more common in males of the same snail species (Lysaght 1941), and males of *Littorina obtusata* had a significantly higher infection rate with larval trematodes than females at one site of the North American Atlantic coast, although females were more abundant (Pohley 1976). Sometimes female snails are more heavily infected with trematodes, for instance *Littorina neritoides* near Plymouth, England. The copepod *Ergasilus auritus* showed a preference for females of the estuarine fish *Gillichthys mirabilis* (Noble *et al.* 1963).

Age of host More common than preference for one of the two sexes is a preference for hosts of a certain age, as shown by the following examples of monogeneans. Llewellyn (1962) showed that *Gastrocotyle trachuri* and *Pseudaxine trachuri* are most common on young fish and much less frequent on two and three year old fish (*Trachurus trachurus*) and probably even rarer on still older fish. For *Discocotyle sagittata*, Paling (1965) found an increase of infection intensity and frequency with the age of its host, *Salmo trutta*. Further examples are given by Bychowsky (1957): many *Gyrodactylus* infect 100 per cent of young fish but adults only in exceptional cases; *Protancyrocephalus strelkovsky* infects 100 per cent of young flounder and is practically absent on adults; *Diclybothrium armatum* is not found on young sturgeon, but often infects 40–80 per cent of the adults. *Dactylogyrus skrjabini* infects 25 per cent of Amurean *Hypophthalmichthys* less than a year old with not more than 3 parasites per fish, but 70 per cent of two year olds with up to 20 parasites per fish (Achmerov 1954). *Microcotyle spinicirrus* is not found on *Aplodinotus grunniens* less than one year old, but infects most fish older than two years (Remley 1936, 1942).

Age preferences are also found among many other groups of parasites. The pyramidellid snail, *Odostomia impressa*, parasitizes mainly old oysters along the southern part of the United States Atlantic coast, whereas *O. bisuturalis* parasitizes mainly young oysters in the northern part of that coast (Hopkins 1956c). Parasite burdens with the copepod *Mytilicola* in southwestern England increase with age (Davey and Gee 1976).

Season Some examples of seasonal fluctuations in parasite infections are discussed in Chapter 8. Such fluctuations are common but not universal in cold–temperate environments, but little is known about them in tropical waters.

With regard to Monogenea in temperate waters, some species occur throughout the year while others are restricted to certain seasons. Several freshwater species have been examined particularly well (Gröben 1940, Bychowsky 1957, Paling 1965) and preferences for cold or warm seasons have also been demonstrated experimentally (Bogdanova 1957). Larvae of the marine monogeneans *Gastrocotyle trachuri* and *Pseudaxine trachuri* on *Trachurus trachurus* are common in May. Only adults are found in July and August (Llewellyn 1959).

Food There is a wide range of food utilized by parasites, depending on the species of parasite and the site of infection. Only some examples will be discussed here.

Arme (1976) reviewed feeding in Protozoa and helminths, but most of his examples are non-marine. Data on food and feeding of some crustacean parasites of fish can be found in Kabata (1970). Examples of blood feeders from his book are: the branchiuran *Argulus*, females of the copepod *Sarcotaces*, and larvae of the isopod family Gnathiidae; copepods of the family Lernaeopodidae apparently feed mainly on tissue and mucus, but sometimes probably also on blood of particular body parts, such as the fins. The food may sometimes differ with the sex or the age of the parasite. Thus, according to Brandal *et al.* (1976), the copepod *Lepeophtheirus salmonis* feeds on the blood of its host, the Atlantic salmon *Salmo salar*, but adult females do this to a greater extent than males and postchalimus larvae.

Several studies have shown that the polyopisthocotylean monogeneans feed on blood and also in some instances on small amounts of tissue and mucus (review by Rohde 1975a). Corresponding to blood feeding is a characteristic structure of the intestine, as seen under the electron microscope (Halton 1974, reviews by Rohde 1975a, 1980c). Monopisthocotylean monogeneans, on the other hand, usually do not ingest blood, but live instead on epidermis and associated mucoid secretion (Llewellyn 1954, Halton and Jennings 1965), although in some species blood may be the main component of food (Uspenskaya 1962, Khalil 1970). Of particular interest is the monopisthocotylean, *Trochopus pini*, which has evolved from skin inhabiting and skin feeding ancestors but occupies the gills of the marine fish, *Trigla hirundo*. A variety of techniques failed to provide evidence for blood feeding and it has hence apparently retained the skin feeding habits of its ancestors (Kearn 1971). Even related forms may use different food resources. The monogenean *Amphibdella flavolineata* is a blood feeder, and the related *Amphibdelloides maccallumi* probably feeds on epidermis (Kearn 1963b).

Among the endoparasites, many trematodes ingest food particles in the lumen of the host's intestine, and cestodes absorb food from the lumen of the intestine through their tegument.

Saturation of niches with species

Most ecologists assume that evolution has filled all niche "space" and that, consequently, species are engaged in bitter competition for the resources available. Competition, it is believed, either leads to the complete disappearance of one or some of the competing species, to their competitive exclusion, or it leads to the narrowing down of the niches of all or some of the competitors (fundamental niche becomes realized niche, see page 104). In an evolutionary context, the assumption of niche saturation implies that species numbers either cannot increase further, or that an increase is possible only by a decrease in niche size of species.

Two arguments can be given against saturation of all niches. Firstly, one can argue that species have not yet evolved that are suitable for some vast spaces which are now obviously empty, such as the upper atmosphere which is not reached by birds, the large ice covered areas of the Arctic and Antarctic which are rarely visited by birds or mammals, or the vast spaces of oceanic midwater which have relatively few fish or large invertebrates. However, can the possibility be excluded that organisms will still evolve for these spaces? A biologist working in early geologic history, before land and air were conquered by animals, might well have concluded that the invertebrates and fish in the sea were all that could possibly evolve.

Secondly, one can estimate the number of available niches based on those used in well occupied ecosystems and compare them with the number of extant species. However, most ecosystems are too complex for such an evaluation. It is difficult to judge how many ant species a forest can support, or how many fish species a coral reef can accommodate. Rohde (1976b,c, 1977a,b,c, 1978b,e, 1979c, 1980a,d,f, 1981a), therefore, used the surface and particularly the gills of marine fishes and their parasites for such a study. Such parasites are an ideal model for several reasons:

1. The marine environment is more uniform than freshwater and terrestrial environments;
2. Environmental variability can be further reduced by considering small habitats in the open water, that is, the surface of fish;
3. The surface of fish and particularly the gills are habitats which can be examined accurately and quantitatively in a short time;
4. The distribution of ectoparasites in certain microhabitats can be accurately mapped;
5. Many marine fishes are easily available in large numbers (Rohde 1981).

If, in the course of such a study, fish species are encountered which have a much smaller number of parasite species than other fish of similar size and habits, the conclusion is inevitable that the former are not saturated with parasite species, that niches are empty.

Rohde (1976b,c, 1977b,c, 1978e, 1979c, 1980a,f, 1981a) concluded on the basis of the comparison of many fish species from different latitudes that "only a small proportion of niches available to Monogenea and other ectoparasites of fish are filled . . .". Several hundreds of thousands of niches are available to ectoparasites of fish (marine and freshwater), but only approximately 5,000 ectoparasitic species have been described so far (Rohde 1979c). Although many species remain undescribed, it is highly unlikely that several hundred thousand are still unknown.

A few examples may illustrate the differences between various marine fish species in their degree of saturation with parasite species. At Coffs Harbour, northern New South Wales, Australia, 29 snapper, *Chrysophrus auratus*, 40 bream, *Acanthopagrus australis* and 34 tarwhine, *Rhabdosargus sarba*, were examined for ectoparasites (Roubal, 1979; Armitage, 1980; and personal observations). All fish belong to the same family, Sparidae, and the specimens examined were all adults of about the same size. The first species yielded 17, the second 20 and the third 2 species of ectoparasites. Similarly, differences in infections with endoparasites are great, and generally fish at high latitudes have a much smaller number of parasitic species than fish at low latitudes. Parasite diversity also is greater in the Pacific than in the Atlantic Ocean (page 137). There is no reason to assume that a fish from the cold Atlantic could not accommodate as many parasite species as one from the warm Pacific or at least the cold Pacific. In other words, there can be little doubt that many potential niches on the fish are empty.

More evidence for the availability of empty niches is the apparent scarcity of hyperparasites in the marine environment (see Chapter 4). For example, it has been shown in several cases that ectoparasites of marine fish are hosts to parasites, for example the copepod *Caligus* to the monogenean *Udonella*, but most parasites are without hyperparasites and they represent niches still to be filled (Rohde 1978e). Price (1980) also concluded that many resources remain unexploited.

Causes of niche restriction

The foregoing discussion has shown that all parasites have restricted niches, characterized by certain hosts, microhabitats, macrohabitats, etc. Two questions arise. Firstly, what are the proximate causes of niche restriction or, in other words, which physical and chemical factors determine that a parasite inhabits its niche and not some other? Secondly, what is the biological function or, in other words, what is the ultimate cause, if any, of restricted niches? The following paragraphs are devoted to a discussion of these two questions for some of the important niche dimensions.

Host specificity Frequently, ecological factors contribute to limiting the host range. An example was given by Williams (1961). The tapeworm, *Echeneibothrium maculatum*, infects one host species, the ray, *Raja montagui*. Species of *Raja* often form unispecific and at certain times of the year unisexual shoals, without extensive migrations, which leads to isolation of different species and the reduction of the chances of becoming infected with the parasite species of another host. Pearson (1968) described another example. The trematode, *Paucivitellosus fragilis*, infects the blenny, *Salarias meleagris*, and the mullets, *Mugil cephalus* and *Crenimugil crenilabis*, on the Great Barrier Reef, and the mullet, *Liza argentea*, at Brisbane, eastern Australia. All these fish are grazers and become infected by ingesting the cercariae which live attached to the substratum. Other fish in the same areas, which are not grazers, do not become infected.

In all cases of "ecological host specificity" (see page 105), host specificity is entirely or largely determined by such ecological factors. If ecological barriers are experimentally removed, for instance by keeping hosts in aquaria, such host specificity is often reduced. Thus, Nigrelli (1947) observed that several fish species which were never infected with the monogenean *Benedenia melleni* in nature became infected in aquaria. However, the natural hosts were most heavily infected, indicating that other features in addition to the ecological ones are also involved.

Other factors responsible for host specificity are of several kinds, and they may act at various stages of the infection process or after infection has occurred. Factors effective during the infection process are those which release hatching, direct the infective stages to the host and bring about settlement of the parasites. These factors may be specific (parasites infect only their "correct" host(s)) or they may be non-specific (a range of "correct" and "incorrect" hosts is infected). In the latter case, host specificity is due to mortality of parasites in the wrong hosts occurring after infection.

Examples for both specific and non-specific factors active during infection were discussed in Chapter 5 (pages 61–65). In some species of Monogenea, hatching is induced by a specific factor in the mucus of the host fish, in others the factor is non-specific (for species specificity of fish mucus, see Barry and O'Rourke 1959). Furthermore, in some monogeneans there is an endogenous hatching rhythm adapted to the activity pattern of the host, in others there is no such rhythm. Finally, larvae may be actively attracted by the host and once larvae are in contact with the host, they may, or may not, react to some specific factor produced by it. For instance, Kearn (1973) has shown that most eggs of the monogenean, *Entobdella soleae*, hatch during the first four hours of illumination. The hatching rhythm is apparently an adaptation to the behaviour of the host, *Solea solea*, which is active at night and partly buries itself in the sediment at dawn. During the day it is inactive and an easy target for the parasite's larvae. Larvae also are attracted by some factor in the skin of the host (Kearn 1967). Host specificity of endoparasites in many cases may be due to specific factors of the host which are necessary for hatching of infective stages after ingestion (review by Lackie 1975). For the copepod, *Lepeophtheirus pectoralis*, Boxshall (1976) showed that two strains, one infecting flounder and the other plaice, had rheotactic responses (responses to water currents) ensuring their contact with their host, whereas chemical factors presumably are responsible for settlement on the right host.

Differential mortality after infection leads to host specificity in many cases. For example, species of the monogenean *Gyrodactylus* attach thamselves readily to abnormal hosts, even to frogs, but subsequent development is poor and all worms die after short periods (Lester 1972, Lester and Adams 1974). If the monogenean *Entobdella soleae* is experimentally transferred to wrong hosts, worms become detached after only 24–30 hours, although they survive on glass for 2–6 days (Kearn 1967, 1970b). Non-specific infection followed by differential mortality in the wrong hosts is also probable for certain trematodes (James 1971) and copepods (Boxshall 1976). The factors responsible for mortality have been analysed in some cases. For example, the cestode, *Acanthobothrium quadripartitum*, is a gut parasite of the ray, *Raja naevus*, but not of the closely related *R.*

radiata. In vitro, it survives for more than 24 hours in fresh serum from the natural host, but 80 per cent of the worms are killed in serum from *R. radiata* within 2 hours. It is possible to abolish toxicity of the "wrong" serum by certain procedures (McVicar and Fletcher 1970). Apparently a substance in the serum of the wrong host is responsible for host specificity.

Kearn (1963b) suggested that blood borne antibodies are unlikely to be responsible for host specificity in epidermis feeding monogeneans, because fish epidermis is not vascularized. However, O'Rourke (1961) has shown that some blood antigens may be found in skin mucus, Diconza and Halliday (1971) demonstrated immunoglobulins in the mucus of catfish, *Tachysurus australis*, and Fletcher and Grant (1969) found antibodies similar to those found in serum in the mucus of plaice. According to Nigrelli (1935b), *Epibdella melleni* lives longer in mucus from susceptible fish than in mucus from "immune" fish. Jahn and Kuhn (1932) also thought that a humoral immunity is responsible for rejection of *E. melleni* (see also Nigrelli and Breder 1934, Nigrelli 1937). In the monogenean *Gyrodactylus alexanderi* on *Gasterosteus aculeatus*, on the other hand, a humoral immune response is unlikely, because: (1) previously infected fish do not lose worms at a faster rate than unexposed fish; (2) loss of worms is very rapid; and (3) intramuscular injection of whole fluke antigen conferred no protection (Lester 1972).

Another mechanism which fish may use to reject certain monogeneans is proliferation of gill epidermis (Wunder 1929, Paperna 1963, Putz and Hoffman 1964), or shedding of a layer of mucoid material from the epidermis to which the worms are attached (Lester 1972). Although no experimental evidence is available, it seems possible that such mechanisms are more effective in relation to "wrong" species and thus determine host specificity.

Frequently, morphological factors are responsible or at least partly responsible for host specificity. For instance, Williams (1961) demonstrated that the tapeworm *Echeneibothrium maculatum* lives in the intestine of a single host species, the ray, *Raja montagui*. The bothridia, sucker-like attachment organs, of this tapeworm are perfectly adapted to attaching themselves to folds of the mucosa in *Raja montagui*, but not to the villi of the related *R. naevus*. A similar "lock-and-key" pattern of attachment organs and hosts' mucosa was found in several other species of cestodes but other factors also contribute to specificity (Williams 1966) (Fig. 63).

Microhabitats Thorson (1969) pointed out that the host represents a remarkably complex array of stimuli, few of which have been adequately studied, and some of which may provide cues for

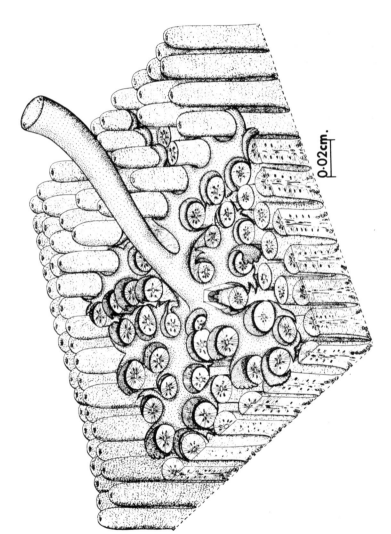

Fig. 63. Anterior end of the cestode *Phyllobothrium pirei* attached to gut mucosa (from Williams *et al.* 1970, after Williams).

migration, settling in the microhabitat, etc. In order to fully understand the factors determining microhabitat specificity of parasites, a detailed knowledge of the morphology and physiology of the various sites is necessary. A detailed discussion cannot be given here, and only a few references on the more important microhabitats of fishes will be quoted.

Accounts of the body surface of fish are by Becker (1942, carp, including defence properties against parasites), van Oosten (1957, structure, chemical composition and function of skin and scales of fishes), Roberts et al. (1971, light and electron microscopic structure of plaice, *Pleuronectes platessa*), Bremer (1972, secretory elements of teleost epidermis), Bullock et al. (1976, histology and histochemistry of whiting, *Merlangius merlangus*) and Mittal et al. (1980, fine structure and histochemistry of *Monopterus cuchia*, with special reference to mucus producing cells). Kearn (1976) reviewed the body surface of fishes as a habitat for parasites. It is of special importance that the epidermis, including the mucus cells, of teleost fish shows mitotic activity (Henrickson 1967, Pickering and Macey 1977, Bullock et al. 1978). The potential food resource for parasites—mucus and cells—is therefore continuously replaced. Furthermore, there is a rich bacterial flora on the skin, gills and in the gut of fish (Horsley 1977), which may be an important though little studied food source for parasites.

Fry (1957) reviewed the function and structure of fish gills, Hughes (1960) studied breathing movements of the jaws and operculum and associated pressure changes in the mouth cavity in 13 species of marine fish, Hughes and Grimstone (1965) made an electron microscopic study of the secondary gill lamellae of *Gadus pollachius*, and Hughes and Iwai (1978) made a morphometric study of the gills of some deep sea fishes. The structure of the gill surface of the freshwater catfish, *Heteropneustes fossilis*, was examined with the scanning microscope by Rajbanshi (1977), and Fernando and Hanek (1976) discussed the gills of fish as a habitat for ectoparasites. An important but older account by Becker (1942) on carp skin and gills includes a discussion of defence mechanisms against parasitic stimuli.

Important papers on the digestive tract of fishes are by Dawes (1930, histology of plaice, *Pleuronectes platessa*), Barrington (1957, review of histology, cytology and function) and Kapoor et al. (1975, review on morphology, histology, cytology, ultrastructure, histo- and cytochemistry, food intake, digestion and absorption). MacKenzie and Gibson (1970) studied physico-chemical condition (pH, osmotic pressure, concentrations of various ions) in the digestive tract of the flounder, *Platichthys flesus*, in relation to site specificity of helminths in it, and Befus and Oodesta (1976) and Williams *et al.* (1970) reviewed the alimentary canal of fish as an environment for helminth parasites.

The reviews by Ulmer (1970) and Crompton (1976) on site selection and finding of parasites show how little is known, and practically no studies of these aspects have been made in marine parasites. Factors which appear to influence site selection and migrations of endoparasites are, according to Crompton (1976), the internal morphology, age, sex and physiological status of the host.

In some species, the infective stages settle in the definite microhabitat and further development occurs there. This is apparently the case in many nematodes inhabiting the alimentary canal. Ex-sheathment and development in certain species were shown to require a stimulus from the host. Although the stimulus itself is not very specific, it acts on larvae of different species at different conditions of redoxpotential and pH, and thus may determine the different microhabitats (Rogers 1957, Sommerville 1957). The studies were made on parasites of terrestrial animals, but marine nematodes probably behave similarly. Identical distributions of larvae and adults, arising from settlement of infective stages in the definite microhabitat, are known from many marine Monogenea and Copepoda (for example *Caligus brevicaudatus* on *Trigla lucerna*, Rohde 1980d), but the causes of site selection are not known.

In other species, larvae migrate to the definite site after infection. This is the case, for instance, in the copepod, *Caligus diaphanus*, on *Trigla lucerna* in the North Sea. Larvae live on the gill filaments but adults occupy the inner wall of the opercular cavity (Rohde 1980d). Larvae are never found in the mouth cavity and adult females never on the gills; differences in their distribution therefore must be due to migration, but stimuli directing the larvae and inducing final settlement have not been studied. Different distributions of larva and adults may also be due to initial random settlement and mortality in abnormal sites, but no direct evidence for this is available.

Site specificity, at least in some species, may be due wholly or partly to physiological factors. Thus, Laurie (1971) showed that two cestodarians, *Gyrocotyle fimbriata* and *G. parvispinosa*, from the ratfish, *Hydrolagus colliei*, in the northeastern Pacific, absorb certain carbohydrates at different rates. This could be an adaptation to different carbohydrate concentrations in different parts of the digestive tract. The first species occupies a more anterior microhabitat than the second (Simmons and Laurie 1972).

In other cases, morphological adaptations are at least partly responsible for site preferences. Thus, three gill parasites of *Seriolella brama* in New Zealand have different microhabitats: the trematode *Syncoelium* lives on the spiny parts and the monogenean *Eurysorchis* on the smooth parts of the gill arches, and the monogenean *Neogrubea* on the gill filaments (Rohde *et al.* 1980). The first species attaches itself by means of a strong ventral sucker which is closed around one or two spines, the second species has an open clamp with a secondarily developed sucker used for attachment to a flat surface, and the third species closes the two valves of its clamps around the filaments. Each species has attachment organs suitable only for its particular microhabitat. Williams (1960, 1961, 1966, 1968a,b) has shown that several cestode species have attachment organs which fit into the pattern of intestinal villi or folds of certain host species like a key into a lock (Fig. 63). However, this is only a precondition rather than the whole reason for site specificity, because adjacent sites have similar surface structures of the intestinal mucosa.

There are many morphological characteristics of various sites which could be used by parasites for "recognizing" their microhabitat, many at the submicroscopic level and revealed only by recent studies. For example, Hunter and Nayudu (1978) demonstrated with the scanning electron microscope that the epidermis of teleost fish has complex "whorled arrangements of microfolds", particularly on the scales, but also on the tail and body fins. No studies have been made on whether parasites use such folds for attachment. If they do, differences of the microstructure may well contribute to determining host and microhabitat specificity.

For ectoparasites, detailed studies of strength and direction of water flow may cast light on factors determining site specificity. With regard to attachment within a microhabitat, it is well known that copepods and monogeneans on the gills are usually attached with their attachment organs directed towards the gill arches and their body parallel to the filaments (see Boxshall 1974 for copepods, Llewellyn 1956 for monogeneans), and the head of copepods on the body surface and in the mouth cavity is generally directed forward and their longitudinal axis is parallel to that of the host (Boxshall 1974).

Geographical range and macrohabitats As pointed out earlier, a clear distinction between the niche dimensions of geographical range and macrohabitat is not always possible. Some factors determining the geographical distribution are probably similar to those responsible for macrohabitat ranges. Often, absence of final or intermediate hosts in certain areas limits a parasite's range, that is factors affecting the geographical distribution of the hosts also affect that of the parasite. The importance of temperature for the geographical distribution of parasites was discussed in Chapter 8 (pages 151–154).

The factors affecting the distribution of marine parasites in different macrohabitats were extensively studied by Russian authors. The review of this work by Polyanski (1961a) shows the following trends for parasites of marine fishes:

1. Chemical factors and particularly salinity affect the fauna of parasites and especially of ectoparasites. Thus, in the Aral Sea, some groups (myxozoans, ciliates, molluscs) are completely absent in the saline part, and monogeneans and trematodes are reduced compared with the freshwater part.
2. Depth is important, and the effect may be either due to pressure or temperature. The importance of temperature is, for instance, indicated by the distribution of the trematode, *Derogenes varicus*, which occurs in surface waters at high latitudes but in deeper cold waters at low latitudes (Manter 1934).
3. The habits and diet of fish largely determine the variety, intensity and frequency of parasite infections. Thus, the type of diet is responsible for the ingestion of certain types of trematode larvae, a long life span of the host permits accumulation of more parasites, a great mobility and a large area covered by the host expose it to a great variety of parasites, gregarious habits may lead to mutual infection, and a large size and age of a fish usually leads to infection with many parasite individuals and species. The Atlantic cod, *Gadus morhua morhua*, migrates over considerable distances, has a long life span (15 years), a wide food range (in the Barents Sea alone more than 200 prey species of fish, benthos and plankton), and consequently a very rich parasite fauna (71 species according to Dollfus 1953). The importance of diet is indicated by differences in the parasite faunas of littoral fish and plankton feeding fish. Typical littoral parasites are the trematodes, *Podocotyle atomon*, *Prosorhynchus squamatus* and *Lepidapedon gadi*, which have coastal invertebrates as intermediate hosts. Typical parasites of plankton feeding fish are larval nematodes of the genus *Anisakis*, trematodes of the family Hemiuridae, etc. Such parasites can be used to study the mode of the host's life.

Russian parasitologists arrived at the following generalizations, based on numerous studies: plankton feeders have relatively few kinds and numbers of parasites and frequency of infection is low, whereas carnivores have many kinds and numbers of parasites and higher frequencies (Dogiel *et al.* 1961, Polyanski 1955). Noble (1973) confirmed this and added that predators feeding on one or few prey species tend to have fewer parasites than those feeding on more prey species.

Ectoparasites are directly exposed to the macroenvironment (of the host) and they must have adaptations to survive in that environment. Thus, Kearn (1962) demonstrated that the monogenean *Entobdella soleae* on the skin of the common European sole, *Solea soleae*, has undulating movements of its body which increase its oxygen supply. The host fish lives near the sea bottom which is poor in oxygen and it spends most of the day buried where even less oxygen is available. "Breathing movements" of the worm are therefore essential. That oxygen is in fact consumed by marine monogeneans was demonstrated for *Diclidophora merlangi* under a variety of experimental conditions (Arme and Fox 1974). Sometimes an ectoparasite is adapted only to part of the macrohabitat range of its host, sometimes to all of it. Both have been demonstrated experimentally for two parasites of marine fish by Nigrelli (1935a). Fish survived in aquaria without apparent discomfort if the salinity was increased, whereas the monogenean *Benedenia melleni* died. On the other hand, the copepod *Ergasilus labracis* survived at salinities ranging from 0.1 to 32 per thousand like its host, *Morone saxatilis*.

Sex and age of host; season Differences between the sexes and seasonal changes in infection with ectoparasites are possibly connected with sexual differences and seasonal changes in the composition of the skin, as demonstrated for the freshwater brown trout, *Salmo trutta*. Pickering (1977) found that the epidermis of both male and female fish undergoes rhythmical changes in thickness during successive spawning cycles, and the male has a significantly thicker epidermis than the female for most of the year. In December and January, the spawning period, the number of mucous epidermal cells in the male drops significantly. Host skin mucus may act as a hatching stimulant for certain marine Monogenea (see page 62, MacDonald 1974) and may explain differences in infection of male and female fish and between different seasons. The importance of temperature for reproduction and occurrence of certain species of freshwater Monogenea was demonstrated by Bogdanova (1957).

Endoparasites may infect the two sexes differently because male and female fish often have different feeding habits (for instance certain mullet in Australia).

Age difference in infection with long lived parasites often may be due to a simple accumulation effect. If the parasites are short lived, other factors must be involved. Thus, Davey and Gee (1976) suggested that large mussels in southwest England are more heavily infected with the copepod *Mytilicola* because they have a higher filtration rate. Furthermore, older hosts usually are larger and present more and larger microhabitats to parasites, and they consume larger quantities of food resulting in a greater probability of becoming infected with orally acquired parasites.

Biological functions of niche restrictions

A discussion of the biological function of niche restriction must proceed from an analysis of biological factors responsible for niche restriction. A critical evaluation of such factors was given by Rohde (1979c, 1980a). They may be of two kinds, extrinsic, due to interspecific effects, or intrinsic, due to intraspecific effects. The most important extrinsic factor is believed to be interspecific competition, but predation, hyperparasitism and reinforcement of reproductive barriers have also been implicated (Rohde 1979c). The only intrinsic factor which has been suggested to lead to niche restriction in parasites is enhancement of the chances to mate (Rohde 1976b,c, 1977a,b,c, 1978e, 1979c, 1980a,d), although intrinsic factors like crowding, intraspecific competition and certain behavioural mechanisms regulating population growth may lead to niche expansion (Rohde 1979c).

Extrinsic factors In the marine environment, predation on parasites could be by cleaner organisms and, to a lesser degree, perhaps by one parasitic species devouring another. However, the section dealing with cleaning symbiosis has shown that there is little evidence for population control of parasites by cleaners (Atkins and Gorlick 1976), and even in habitats with no or few cleaners, densities of ectoparasitic populations are usually low. Even if it could be shown that cleaners affect many populations of ectoparasites significantly, they could have no effect on the endoparasites which constitute the majority of parasitic species.

That one parasitic species may prey on another has been shown for some species of larval trematodes in freshwater snails (Lie et al. 1968) but it is not common and has so far not been found in marine parasites (Rohde, 1981b).

Marine hyperparasites are relatively rare and any regulating effect due to them would be slight (see Chapter 4).

The only remaining extrinsic factors which may play a significant role in restricting niches of marine parasites, then, are interspecific competition and reinforcement of reproductive barriers. Generally, competition is thought to be by far the most important of the two.

Rohde (1979c) and Price (1980) have reviewed the evidence for interspecific competition among parasites and concluded that its significance has been overestimated. Holmes (1973), on the other hand, argued that competition is the main factor determining community structure in parasites, partly as a presently active agent, but more importantly as an agent active in the evolutionary past. According to him, some parasitic species show interactive site segregation, that is their niches are compressed by competing species (fundamental niche becomes realized niche, see page 104); most parasite species, however, show selective site segregation, that is their niche is not changed due to competing species (fundamental niche is equal to realized niche). This is explained by the hypothesis that selection has removed characters which led to competition in the past. In other words: in the course of evolution some species have either disappeared altogether due to "competitive exclusion", or the original realized niches of species have become their extant fundamental ones. Because of the stable niches of most parasite species which are little affected by other species, parasites, according to Holmes (1973), live in mature communities with a long evolutionary history.

Interaction among parasite species has indeed rarely been observed. Halvörsen (1976) gave some examples mainly from non-marine forms, MacKenzie and Gibson (1970) demonstrated it in two species of the nematode *Cucullanus*, and Holmes (1971) in two species of blood flukes in the fish, *Sebastes* spp. In most cases, frequencies and intensities of infection and sites of infection remain unaltered in the presence of other species (Fenchel 1965, 1966 for ciliates on bivalves; Lauckner 1971, for trematode larvae of the bivalves, *Cerastoderma edule* and *C. lamarcki*). Frequently there even is a positive association, that is host individuals have simultaneous infections with several species more often and/or more densely than expected if infections were random (Noble *et al.* 1963, Bortone *et al.* 1978).

Even where interaction between species occurs, it is by no means certain that competition is responsible. For instance, segregation of the two species of *Cucullanus* mentioned above could as well be the result of reinforcement of reproductive barriers. And even if competition could be shown to occur, it does not follow that it is an important evolutionary agent. It is difficult to provide direct evidence for the evolutionary effect of competition because evolution is slow relative to the duration of research projects investigating it. The only direct evidence comes from the gradual spreading of species and replacement of others, and from introduction of species into the habitats of others by man. Because so few cases of direct evidence have been examined and none for parasites, indirect evidence is used, and the most important evidence of this kind is character displacement. Two related species often show the greatest differences in that part of their area of distribution where they overlap (Brown and Wilson 1956). Most authors assume, without critically examining the data, that character displacement must be the result of competition. However, as already pointed out by Brown and Wilson (1956) and especially stressed by Miller (1967), many cases of character displacement can as well or better be explained by reinforcement of reproductive barriers.

Ectoparasites of fish are an excellent model for examining the relative importance of interspecific competition and reinforcement of reproductive barriers, because related (congeneric) and unrelated species using the same limiting resources can be compared. Competition for these resources should occur irrespective of whether species are related or not (although many authors uncritically assume that competition should be strongest between related species, supposedly because of their similar requirements). The factor limiting populations of parasites on the gills of marine fish appears to be space, and possibly food for those copepods and monogeneans which feed on mucus and epithelium of the skin with its limited capacity for regeneration (Llewellyn 1966). Blood, the food for many monogeneans and copepods, is amply available and not a limiting factor.

Rohde (1980d) found that in several species of Monogenea and Copepoda on the gills of marine fishes in the North Sea, separation was clearest in congeneric species pairs. The monogeneans *Kuhnia scombri*—*Kuhnia* sp. on *Scomber scombrus*, the copepods *Caligus diaphanus*—*C. brevicaudatus* and *Neobrachiella impudica*—*N. bispinosa* on *Trigla lucerna*, and possibly the copepods *Clavella iadda*—*C. sciatherica* on *Gadus morhua* all had practically non-overlapping microhabitats. The same species, on the other hand, were often in contact with unrelated species, that is the two species of *Kuhnia* with cysts and the copepod *Caligus pelamydis*, and the copepod *Caligus diaphanus* with the copepods *Neobrachiella impudica* and *Lernentoma asellina*. This seems to indicate that reinforcement of reproductive barriers rather than interspecific competition is responsible for the segregation. However, as pointed out by Rohde (1980a,d) there is often much overlap even between congeneric species (for instance in two species of the monogenean *Diplectanum* on the sea bass, *Dicentrarchus labrax*, studied by Lambert and Maillard 1975). More studies are necessary to evaluate the relative importance of the two factors for species segregation.

Neither of the two factors, competition or reinforcement of

reproductive barriers, can be responsible for niche restriction if niches are restricted even when interacting species are absent. As shown on pages 116–118, only few of the niches available to ectoparasites of fishes (and to other parasites) are utilized, particularly at high latitudes. Nevertheless, microhabitats in communities with few species are as narrow or as wide as in communities with many species (Rohde 1981a, pages 144–147). Reasons are that, in the former, species do not expand into adjacent empty microhabitats and that, in the latter, species show much overlap and do not contract because of the presence of other species.

Geological evidence indicates that species numbers of free-living animals at high latitudes have been smaller than those at low latitudes for hundreds of millions of years. The same can be assumed for parasites. Competition and reinforcement of reproductive barriers in the past, therefore, cannot have been responsible for restricting niches at least at high latitudes.

It is often argued that even if organisms have identical or similar requirements along certain niche dimensions, they may be segregated along another dimension, and exploitation of different food resources as indicated by different types of mouth organs is thought to be of foremost importance. However, the parasites discussed above have identical or similar food requirements, but their feeding organs are of different size and structure (Rohde 1979a,c). This shows that such differences often may be purely coincidental, without any consequences for niche segregation.

In summary, we must conclude that competition between parasite species occurs, but its evolutionary significance is probably insignificant, largely because so many potential niches remain unoccupied.

Price (1980) also concluded after a critical examination of the data from many studies that competition is not important as a force in parasite evolution. "Where competition exists, it may frequently result in nonequilibrium transient competitive displacement, a disruptive influence in community development." Kennedy (1977) considered non-equilibrium conditions for parasites of fish possible.

Intrinsic factors Intrinsic factors sometimes bring about expansion of an organism's niche width. One such factor is complex behavioural mechanisms regulating population growth (Wynne-Edwards 1962, Coulson 1971, Way and Cammell 1971). However, these mechanisms cannot be expected to occur in lower organisms to which practically all parasites belong. Crowding, or abnormally high population density, on the other hand, repeatedly has been shown to lead to niche expansion in parasites, probably as a result of increased intraspecific competition. Anderson (1974) demonstrated enlarged microhabitats in high density populations of the monogenean, *Diplozoon paradoxum*, on the freshwater fish *Abramis brama*, and further examples from other parasitic groups were given by Crompton (1973).

Enhancement of the chances of mating is the only intrinsic factor which has been suggested to be responsible for niche restriction in parasites (Rohde 1976b,c, 1977c). Ktari (1971) already had noted that narrow microhabitats increase the chances of Monogenea of mating, but he did not draw any evolutionary or ecological conclusions.

Mating is of great significance to most species of parasites. This is clearly indicated, for instance, by the great complexity of copulatory structures in many groups. Its importance is self-evident for all bisexual forms, but has also been demonstrated for many hermaphrodites. Hermaphroditic parasites may be obligatory or facultative cross-fertilizers. *Lobatostoma manteri*, an aspidogastrid trematode in the intestine of the marine fish, *Trachinotus blochi*, on the Great Barrier Reef, for instance, is an obligatory cross-fertilizer. In single infections, normal looking eggs are produced, but cleavage does not proceed to the blastula stage (Rohde 1973). On the other hand, the trematode, *Zygocotyle lunata*, can successfully fertilize itself, as indicated by the fact that infections with single larvae resulted in adult worms capable of producing viable eggs (Bacha 1966). Bacha (1966) and Nollen (1968) gave more examples for both cases. Some species, although capable of successful self-fertilization, nevertheless produce a greater number of successful offspring when cross-fertilized.

In the Monogenea, Kearn (1970a) described mating in the skin parasite *Entobdella soleae* on the European sole, *Solea solea*. According to him, there is no evidence for self-fertilization in this species. In the viviparous monogenean *Gyrodactylus alexanderi*, the first daughter animal is produced without cross-fertilization, although cross-fertilization may occur for production of the second embryo (Lester and Adams 1974).

Self-fertilization hermaphroditism is common among tapeworms (Baylis 1947 and Sandars 1957 cited by Williams 1966, Williams 1966), although cross-fertilization may occur.

Parasites may encounter difficulties in mating because frequencies and intensities of infection with most parasites are usually low. Examples for ectoparasites of marine fish were given by Rohde (1979c, 1980d). Most exceptions are species parasitizing host populations of abnormally high densities, usually due to human influence, and species which can multiply within or on the host.

Examples of the former are discussed on pages 97–98. Examples of the latter are species of the monogenean family Gyrodactylidae which are viviparous, the larvae developing without leaving the host. Of 25 specimens of whiting, *Sillago* sp., from northern New South Wales, Australia, for instance, only 3 were infected with a gyrodactylid, but infection intensities were high: 380, 365 and 25 specimens, respectively, apparently as a result of multiplying on one host individual.

Low density populations of parasites would hardly have a chance to establish contact between mating partners if their microhabitats were not drastically reduced in size. Rohde (1979c) described an example. He found a total of 11 attached specimens of the monogenean, *Allopseudaxinoides vagans*, on eight skipjack tuna, *Katsuwonus pelamis*, in New Caledonia; 31 fish were not infected.

> All specimens were recovered from the first few (less than the first 50) gill filaments of the left and right gills nos. 3 and 4, i.e., from less than 200 filaments. The total number of filaments of all four gills, left and right, is approximately 2,500. Parasites of the group to which the species belongs are known to move around very little, if at all, and they cannot be expected to bridge more than a few gill filaments by extending their small, few-millimeter long bodies. Given the low frequency and intensity of infection, the chances to mate would be nil without the extreme microhabitat restriction, which increases the probability of contact more than twelvefold.

Rohde (1976b) used examples of copepods and monogeneans on the Spanish mackerel, *Scomberomorus commerson*, to show that microhabitat restriction indeed leads to increased intraspecific contact.

The "mating hypothesis" of niche restriction suggests that species which have other means of establishing intraspecific contact, such as good locomotory ability or ability to establish large populations, have less restricted microhabitats (Rohde 1979c, 1980d). Asexual stages like cysts or larvae also can be expected to have large microhabitats, although they may show site preferences because of requirements not connected with mating, for example a suitable substratum for attachment and feeding (Rohde 1979c).

Examples for all three cases were given by Rohde (1979c, 1980d), but more studies are necessary.

Among Monogenea, locomotion has been studied in a number of species (Izjumova 1953, Bychowsky 1957, Kearn 1963a, 1970b, 1971, Lester 1972). Monogenea with large posterior suckers on the skin of fish frequently change their site, and they may inhabit large areas of the body surface. Sessile or semisessile Monogenea on the gills, on the other hand, often have very narrow microhabitats, as predicted by the mating hypothesis (Rohde 1979c, 1980d). Of the parasites on 9 species of marine fish from the North Sea and Papua New Guinea, the copepods, *Caligus* spp., have the best locomotory ability, and they also have the largest microhabitats. *Caligus pelamydis* occurs in the mouth cavity, in external folds of the head, and on the gills of *Scomber scombrus*, *Caligus diaphanus* inhabits gills and mouth cavity, and *C. brevicaudatus* most parts of the body surface of *Trigla lucerna*, whereas the sessile *Neobrachiella impudica*, *N. bispinosa* and *Lernentoma asellina* on the same fish have much more restricted sites.

However, other factors beside enhancement of intraspecific contact may explain at least part of the differences. Thus, Llewellyn (1964) concluded that skin dwelling Monogenea move around frequently because the rate of skin erosion due to feeding exceeds the rate of regeneration and the parasite is forced to find a new feeding site. Even those monopisthocotylean monogeneans which have invaded the gill chamber and still feed on epidermis, have kept their locomotory ability (for instance *Trochopus pini*; Kearn 1971). Polyopisthocotylean monogeneans, on the other hand, feed on blood primarily taken from the gills, and they are sedentary because their food resource is not made more accessible by locomotion.

With regard to the second and third points, larger microhabitats in high density populations and asexual stages, Rohde (1979c, 1980a,d) has shown that cysts on four species of marine fish and a small monogenean occurring in large numbers are randomly distributed over the gills, whereas several species of ectoparasites occurring in small numbers and in the sexually reproducing stage have clearly restricted sites (Fig. 64). Among the latter, the monogenean, *Kuhnia scombri*, on *Scomber scombrus* occurs in relatively large numbers and, correspondingly, has a much larger microhabitat than *Kuhnia* sp. on the same host which is rare and restricted to the pleurobranchs (Fig. 62).

A less restricted microhabitat of juveniles than of adults was shown for the parasitic prosobranch snail, *Thyca crystallina*, on the starfish *Linckia laevigata*. The smallest snails tend to settle on the upper surface of the distal parts of the arms and are randomly oriented, whereas the largest snails, all females, face the starfish mouth in a small area on the right side of the ambulacral groove on the surface of the oral arm. Dwarf males are attached to most of the large females. Juveniles apparently slowly move to the final site, but adults are probably permanently attached (Elder 1979).

In summary, it appears that the mating hypothesis can explain many cases of niche restriction at least partly; enhancement of the

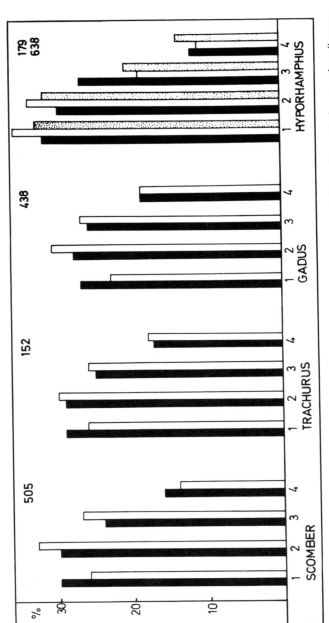

Fig. 64. Distribution of cysts on gills nos 1-4 of *Scomber scombrus*, *Trachurus trachurus*, *Gadus morhus*, and of cysts and small monogeneans on gills of *Hyporhamphus quoyi*. Black columns: area of gill filaments, in percentage of total area of all four gills. White columns: number of cysts as percentage of total numbers on all four gills. Stippled columns: number of small monogeneans as percentage of total numbers on all four gills. Number of cysts and monogeneans above columns. Abscissa: gills nos 1-4 of four fish species. Ordinate: area or number of parasites per gill as percentage of total (after Rohde 1980d).

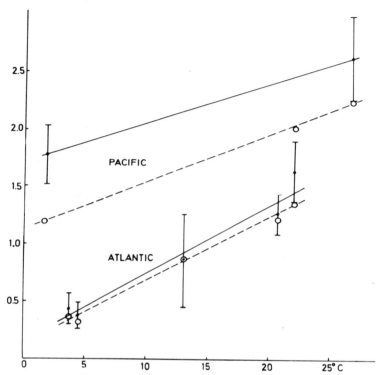

Fig. 65. Relative species diversity (average number of species of parasites per host species) of monogenean gill parasites of teleost fish in the Pacific and Atlantic Oceans. Abscissa: approximate menas of annual sea surface temperature ranges at various localities. Ordinate: mean numbers of monogenean species per host species. ○ = total number of monogenean species/total number of host species examined at each locality. ● = ○, but monogenean species occurring in x host species counted x times (after Rohde 1980f).

chances of contacting a mating partner may be an important biological function of niche restriction in parasites. It has to be kept in mind, however, that parasites often may have restricted microhabitats simply because in an environment composed of many topographically different sites, organisms have to be very specialized in order to survive (Price 1980), and special adaptations to one site may exclude colonization of another site. More studies are needed to evaluate the relative importance of the mating hypothesis.

CHAPTER 8

Characteristics of Parasite Faunas of Different Seas

Zoogeographical regions

Lebedev (1969b) examined 216 genera of Monogenea and 420 genera of Digenea from marine fish and distinguished 10 zoogeographical regions differing in the composition of their Monogenea and Digenea faunas. These regions are:
1. *Western Tropical Atlantic:* Caribbean, Gulf of Mexico, eastern coastal Florida;
2. *Northern Atlantic:* Gulf of Tschesapinsk, coastal Greenland, Britain, Scandinavia, Barents Sea, North Sea, Baltic Sea;
3. *Mediterranean–Atlantic:* Mediterranean, Black Sea, Atlantic coastal Spain and Morocco;
4. *Red Sea:* Red Sea and Gulf of Aden:
5. *Indian:* coastal India, Ceylon, Bay of Bengal and Arabian Sea;
6. *Sunda-Malayan:* South China Sea, Sulu Sea, Celebes Sea, Java Sea, Flores Sea, Gulf of Siam and coastal Philippines;
7. *Japan:* East China Sea, Yellow Sea, Sea of Japan to 40°N, Inland Sea of Japan, Pacific coastal Japan, Ryukyu Archipelago;
8. *New Zealand–Australia:* Great Australian Bight, eastern coastal Australia, Tasman Sea, New Zealand Shelf, New Hebrides and Fiji;
9. *Eastern Tropical Pacific:* Pacific coastal America between 32°N and 10°S, Revilla Gigedo and Galapagos Islands;
10. *North Pacific:* north of Regions of Japan and Eastern Tropical Pacific.

The parasites of many seas have not been examined sufficiently and large areas are therefore not included in Lebedev's scheme. Furthermore, other groups of parasites need to be examined before a general and comprehensive scheme can be established and compared with those based on the study of free-living animals (for example Ekman 1953).

Differences in the parasite faunas of the Atlantic and Indo-Pacific Oceans

The two ocean systems differ in the number of species of parasites as well as in the composition of the parasite faunas.

Parukhin (1975) examined 6,500 individuals of 200 fish species belonging to 100 families from various zoogeographical regions and concluded that species numbers of nematodes are smaller in the southern Atlantic than in the Indian and Pacific Oceans. Data for Monogenea are represented in Fig. 65. The data base is still small, but it seems that the Indo-Pacific has generally more species of parasites than the Atlantic Ocean. This also is true of most groups of free-living organisms (Fig. 66).

With regard to species composition, evaluation of many surveys shows that many species in the Indo-Pacific are different from those in the Atlantic. Lebedev (1969b) calculated the number of genera of trematodes and monogeneans which are endemic to the Atlantic or the Indo-Pacific, and those which both oceans have in common (Table 5). Many genera have not yet been described, and the data are thus far from complete. Nevertheless, it is clear that endemicity is greater in the Indo-Pacific than in the Atlantic. As expected, endemicity of species is much greater than that of genera. Whereas 56 of 96 genera of Monogenea and 136 of 224 genera of trematodes are common to both oceans, Manter (1940a, 1947, 1955) found that of 147 shallow water species at Tortugas, Florida, only 26 species (17.6 per cent) occurred also in the Pacific, and of the 62 species of trematodes found on the Atlantic side of Central America (Bimini), and the 31 species found in the Gulf of Panama by Sogandares-Bernal (1959), only 5 species occurred in both regions. Nevertheless, according to the latter author, until 1959 the impressive number of 41 trematode species had been found to be common to the American Atlantic and Pacific. In Atlantic and Pacific regions further apart than the Caribbean and the Pacific Central American coast, the number of shared trematode species is much smaller. Thus, of the approximately 304 known trematode species from Japan, only 10 are also known from Florida, 10 from the Mediterranean, 9 from the European Atlantic, 5 from the American Atlantic and 5 from Bermuda (Manter 1955). However, Japanese waters share similarly few trematode species with other parts of the Pacific Ocean. Only 14 of the 304 Japanese species are also known from Celebes, 9 from New Zealand, 6 from the Red Sea and 6 from the American Pacific (Manter 1955). Several other examples can be given, but it may suffice to give details of the data for the European North Atlantic. Of

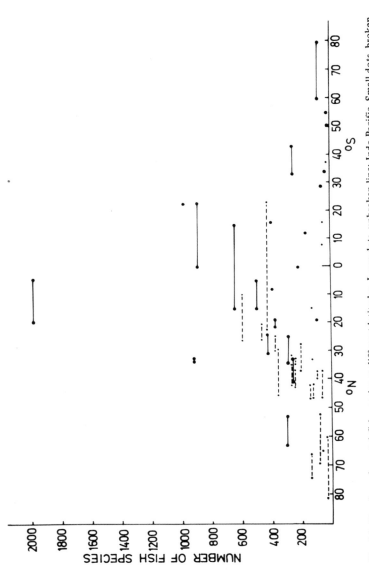

Fig. 66. Number of coastal fish species at different latitudes. Large dots, unbroken line: Indo-Pacific. Small dots, broken line: Atlantic (after Rohde 1977b). Abscissa: latitudes.

Table 5. Comparison of the fauna of Monogenea and trematodes of marine fishes in the Atlantic and Indo-Pacific. Data from Lebedev (1969b).

	Number of genera		Atlantic % of all genera known in Atlantic		% of all genera known in all oceans		Number of genera		Indo-Pacific % of all genera known in Indo-Pacific		% of all genera known in all oceans	
	Mon.	Trem.	Mon.	Trem.	Mon.	Trem.	Mon.	Trem.	Mon.	Trem.	Mon.	Trem.
Total	96	224	—	—	46	55	170*	318	—	—	81	79
Only in 1 region	40	78	41	35	18	19	114	182	67	57	52	45
In both regions	56	136	—	—	27	34	56	136	—	—	27	34

* Arctic, Antarctic and doubtful genera not included.

approximately 72 species known from British waters, 9 are also known from Japan, 6 from New Zealand or Tasmania, 21 from the Mediterranean, 10 from the American Atlantic, 2 from Florida and 2 from Bermuda (Manter 1955).

The number of helminths found both in cold north Atlantic and cold north Pacific waters is surprisingly high (for a review of early findings see Polyanski 1961b, and Dogiel 1964). According to Strelkov (1960), eastern Kamchatka has 51 known species of trematodes, cestodes, nematodes and acanthocephalans, and the Barents Sea has 70. Common to both are 34 species. Of the 57 species found in the White Sea, 26 species have been recorded also from eastern Kamchatka. This is similar to the proportion of helminth species found in adjacent Pacific waters. Of the 51 species from eastern Kamchatka, 35 are also known from the north west part of the Sea of Japan and 26 from the South Kuril Islands. A similarly high proportion of endoparasitic helminths shared by northern Pacific and Atlantic waters was found by Zhukov (1960). Of 87 species from the Sea of Japan, 35 are also known from the Barents Sea, 24 from the White Sea, 29 from the northeast part of North America, and 2 from the southeast part of North America. Again, this is similar to the proportion of species found in adjacent Pacific waters. Thirty-six of the 87 species are also known from the South Kuril Islands, 13 from the northwest part of North America, and 2 from the southwest part of North America.

Latitudinal gradients in species diversity of hosts and parasites

Most groups of animals and plants have more species in warm than in cold environments (review by Rohde 1978d,e). Such gradients are known from fossil communities 270 million years old, and probably they have been in existence for much if not all of evolutionary history. Figure 66 shows the diversity gradients in marine fishes. Data for parasites are much less complete, but distinct diversity gradients can be recognized in those groups which have been studied best, that is in trematode genera of marine fishes (Table 6) and in Monogenea species of marine teleost fishes. For trematode species, latitudinal differences are not clear (Table 7), possibly because some of the surveys are very small. In Monogenea, there is not only an increase in absolute numbers of species towards the equator, expected because of the greater number of host species, but also a *relative* increase. In the tropics there are more species of Monogenea per fish species than in cold–temperate seas (Fig. 65). Data for tropical seas are less complete

Table 6. Latitudinal gradients in numbers of monogenean and trematode genera of commercial marine fishes. Data from Lebedev (1969b).

Biogeographical zone	Number of species of commercial fish	Number of genera of:	
		Monogenea	Digenea
Arctic	approx. 20	15	12
North-boreal	40–50	20	50
South-boreal	> 50	50	160
Tropical	100–120	95	300
North-antiboreal	40–50	40	80
South-antiboreal plus Antarctic	approx. 20	3	10

Note: parasites of tropical fish are much less well known than those of fish from higher latitudes and, since 1969 when this table was composed, many new genera of Digenea and Monogenea have been described from tropical oceans.

than those for cold seas and thus further studies will almost certainly show that the gradients are even more distinct. Data for trematodes and Monogenea are not yet sufficient to show whether there are differences between northern and southern cold seas. But such differences may well exist, as indicated by Parukhin's (1975) finding, based on the examination of 6,500 individuals of 200 fish species (100 families), that the nematode fauna is poorer in the southern than in the northern hemisphere. Szidat (1961) found that the parasite fauna of marine fish is much poorer in the southwestern Atlantic than that in the north Atlantic and Pacific, and there is a distinctly smaller number of helminths of pinnipeds and cetaceans in the southern than in the northern hemisphere (Table 8). The latter group is one of the few examples of reduced diversity in warm waters. Delyamure (cited by Dogiel 1964) explained this by the reduced density of host populations in the tropics leading to less favourable conditions for the development of a rich parasite fauna. An alternative explanation is that hosts and their parasites have originated in cold (boreal–arctic) waters and had more time to acquire parasites in that environment (see pages 154–155).

Latitudinal gradients in frequencies and intensities of infection

Rohde (1977b) compared frequencies (percentage of fish individuals infected) and intensities of infection (number of parasite individuals per host individual) of Monogenea at different latitudes and found that frequencies are greater in tropical seas, whereas intensities vary

Table 7. Number of species of adult trematodes of marine fishes in different regions.

Locality	Number of fish specimens examined	Number of fish species examined	Number of trematode species found	Author
Gulf of Panama and Bimini, West Indies 10–26°N	484	209	87	Sogandares-Bernal (1959)
Red Sea, 12–30°N			44	Manter (1955)
Hawaii, 20°N	approx. 2,000	144	314	Yamaguti (1970)
Tortugas, Florida, 25°N	2,039	237	189	Manter (1947)
Japan, 30–40°N	Many 1,000		299 304	Manter (1947) Manter (1955)
Mediterranean, 31–45°N			107	Manter (1955)
New Zealand, 40–45°S	239	58	66	Manter (1954a, 1955)
South Kuril Islands, 45°N	664	55	44	Zhukov (1960)
Black Sea, 45°N	439	34	36	Osmanov (1940)
British Isles, 50–60°N	Many 1,000		59 77	Manter (1947), Nicoll (1914) and Dawes (1946) Manter (1955)
Kamchatka, 55°N	385	34	25	Strelkov (1960)
White Sea, 65°N	1,376	33	23	Shulman & Shulman-Albova (1953)
Barents Sea, 70°N	1,003	47	35	Polyanski (1966)

Table 8. Helminths species of pinnipeds and cetaceans at different latitudes. Data from Dogiel (1964).

	Trematodes	Cestodes	Nematodes	Acantho-cephalans	Total
Arctic	6	7	9	3	25
Boreal	30	24	31	9	94
Tropical	7	2	18	4	31
Antiboreal	0	9	21	8	38
Antarctic	1	7	5	1	14

greatly and show no definite trends. His results were based on his own data from the Great Barrier Reef and on surveys of other authors. More up to date data from the Pacific and Atlantic oceans have supported these conclusions (Rohde 1981). In both the Pacific and Atlantic, frequencies of infection with individual species of Monogenea increase towards the equator (Fig. 67), whereas intensities of infection are more or less the same at all latitudes.

No such comparisons have been made for other groups of parasites.

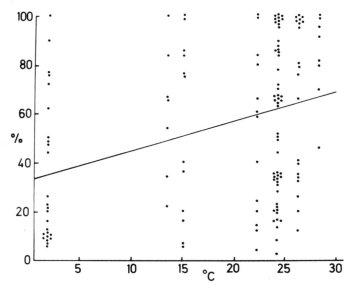

Fig. 67. Frequencies of infection (percentage (1) with Monogenea on the gills of marine teleosts at different latitudes in the Pacific as characterized by means of annual sea surface temperature ranges. Fish species without gill Monogenea not included (after Rohde 1981a).

Latitudinal gradients in niche width

One might expect that, corresponding to the increase in species numbers towards the tropics, there is a narrowing in niche width (where niche is defined as the total of the relations of an organism to its environment, see Chapter 7). In other words, more species can coexist because their niches are smaller.

It is impossible to measure all the parameters determining the niche of a parasitic species (Rohde 1979c, 1981a), but only a few are usually of importance, the most important ones being food, host range and microhabitat. Data are available only for parasites of marine fish, and these will be used to examine the question of whether there are latitudinal gradients in width of "fundamental" niches (see page 104) correlated with the diversity gradients.

Host range It is convenient to distinguish between host range and host specificity. Host range is the number of host species for a certain parasite species irrespective of how frequently and how strongly the various host species are infected, whereas host specificity takes into consideration not only the number of host species, but also the relative preference for each of these. A convenient measure of host specificity is the specificity index proposed by Rohde (1980e) (see page 107).

Data from many surveys show that Digenea of marine fish have, as perhaps expected, more restricted host ranges in warm than in cold seas. Monogenea, on the other hand, have similar host ranges at all latitudes (Fig. 68). Host specificity is similarly high at all latitudes for

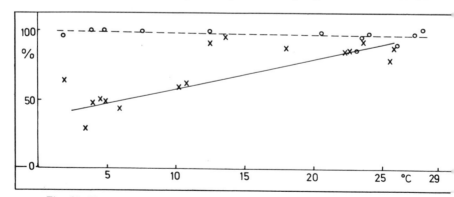

Fig. 68. Host ranges of marine Monogenea (o) and trematodes (x) at different latitudes as characterized by means of annual sea surface temperature ranges. Ordinate: percentage of species on or in one or two host species (after Rohde 1980a).

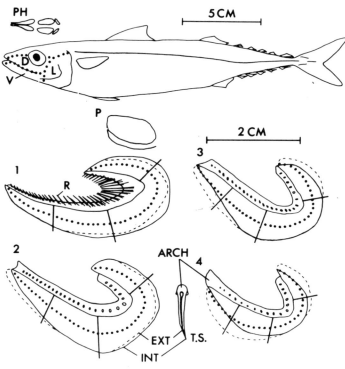

Fig. 69. Diagram of microhabitats on the gills and in the mouth cavity of fish. ARCH – gill arch; D – dorsal mouth cavity; EXT – external gill filament; INT – internal gill filament; L – lateral mouth cavity; P – pleurobranch; PH – pharyngeal plates; R – gill raker; T.S. – cross section through gill; V – ventral mouth cavity; 1-4 – gills nos 1-4. The surface area of the external gill filaments is indicated by an interrupted line, that of the internal filaments by a continuous line, the border between basal and distal zone of the gill filaments by a dotted line. The maximum number of preferred microhabitats is 1 dorsal mouth cavity + 1 ventral mouth cavity + 1 lateral mouth cavity + 1 pleurobranch + (1 gill arch x 4 gills x 4 longitudinal quarters x 2 internal or external surface) + (1 gill filament x 4 gills x 4 longitudinal quarters x 2 external or internal surface x 2 basal or distal half of filaments) = 1 + 1 + 1 + 1 + 32 + 64 = 100 (after Rohde 1981a).

both groups (see page 109) the reason being that, although Digenea infect more fish species in cold seas, even there they show clear preferences for one or a few host species.

Microhabitat Data are available only for ectoparasites of marine fish. Their microhabitats can be accurately measured and counted (Fig. 69). Figure 70 shows the number of preferred microhabitats on

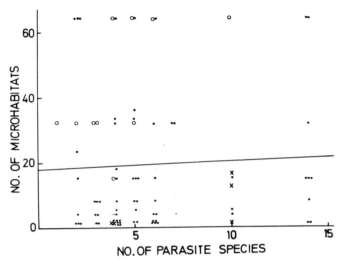

Fig. 70. Numbers of microhabitats per species of parasite on the gills and in the mouth cavity (see Fig. 69) of species of fish with different numbers of parasite species (after Rohde 1981). • = copepods and monogeneans, ○ = cysts, x = didymozoons.

fish species with different numbers of ectoparasitic species. There is no correlation between number of microhabitats and number of parasitic species. Reasons are that species in communities with few species do not expand into empty "niche space" and, conversely, that in multispecies communities there is often much overlap.

Food The third niche dimension considered, food, has been little studied. With regard to the Monogenea, for which most data are available, a considerable number of electron microscopic and other studies has shown that the polyopisthocotylean monogeneans are obligatory blood feeders, whereas most monopisthocotyleans feed on mucus and epithelial cells. Various extensive surveys show that there is a great variability in the ratio Polyopisthocotylea/Monopisthocotylea calculated from the data of the following authors: Hawaii 49/97 (Yamaguti 1968); Gulf of Mexico 41/34 (Hargis 1957); Norway 16/12 (Brinkmann 1952); White Sea 1/10 (Shulman and Shulman-Albova 1953, data corrected by Rohde 1978f); Barents Sea 6/14 (Polyanski 1955); Chukotsk

Peninsula 0/24 (Zhukov 1960, data corrected by Rohde 1978f). No latitudinal trends can be recognized.

In summary, the data available so far indicate that width of the fundamental niche of parasites is not affected by the number of species in the community, that is that interspecific effects, competition and reinforcement of reproductive barriers (see page 127), are probably not of great importance.

Fluctuations of parasite infections in cold and warm seas

Most studies on seasonal fluctuations in population size of parasites have been made in cold–temperate seas. In spite of a distinct seasonality indicated by temperature changes, fluctuations of parasite populations are by no means universal. According to Kennedy (1975), many fish protozoans are apparently not seasonal, whereas Monogenea appear to be almost invariably seasonal. Even populations of the same parasite may differ in this respect. Thus, the copepod parasitic in mussels, *Mytilicola intestinalis*, undergoes seasonal fluctuations in Ireland (Grainger 1951), but not on the North American west coast, where Katkansky *et al.* (1967) made observations at five stations in Washington, Oregon and California.

The monogeneans, *Gastrocotyle trachuri* and *Pseudaxine trachuri*, are examples of clear seasonality of parasites in cold–temperate seas. Their larvae are common in May, and adults only occur in July and August (Llewellyn 1959). Several parasitic species of mullet, *Mugil cephalus*, and mummichog, *Fundulus heteroclitus*, in Georgia, U.S.A., studied by Rawson (1973) are also clearly seasonal. Pohley (1976) found different seasonal variations of several larval trematodes in three littoral snails, *Littorina* spp., on the northeast coast of the U.S.A.

No seasonal fluctuations were found by Noble *et al.* (1963) in four parasite species infecting the estuarine fish, *Gillichthys mirabilis*, in spite of the distinct seasonal changes in water temperatures and salinities (1,112 fish examined over a 3 year period).

The considerable variability in seasonality in cold–temperate waters indicates that a great number of studies of different groups would be necessary to show latitudinal differences. But very few studies in the tropics and subtropics have been made. Rohde and Sandland (1973) and Rohde (1981b) examined the snails *Planaxis sulcatus* and *Cerithium moniliferum* at Heron Island, Great Barrier Reef, for infections with digenetic trematodes and the aspidogastrean *Lobatostoma manteri* for approximately two years and found no seasonality. Infection levels remained more or less constant except for the aspidogastrean. Cannon (1978b, 1979), in *Cerithium moniliferum* at the same locality, found 11 species of cercariae, none of which showed a seasonal pattern of incidence, although some showed considerable irregular fluctuations. Statistical analyses to show the significance of these fluctuations were not made.

Differences between parasite faunas in shallow and deep water

There are few studies on parasites of deep water fishes and most deal with parasites of one or a few fish species only (Dienske 1968, Toman 1973). Gusev (1957), Collard (1970), Noble and Collard (1970) and Campbell *et al.* (1980) made comprehensive surveys, and Noble (1973) summarized the knowledge then available, comparing parasites of shallow and deep water fishes.

In general, pelagic deep water fish have few adult parasites. Collard (1970) arrived at this conclusion on the basis of the examination of 1,122 fish of 44 species and 13 families from depths of 100–4000 m caught mainly off California and Mexico. According to him, there were relatively few adult metazoan parasites, but relatively more larvae. Female fish were generally more heavily infected and some parasites showed seasonal fluctuations, but apparently only those which have seasonal intermediate hosts. Noble and Collard (1970) examined 594 fish of 54 species and 43 genera for protozoan parasites, and 1,087 fish of about 42 species for metazoan parasites and commensals. Most fish were from the "midwater" (100–800 m) off southern California and Mexico, some from the Peru–Chile Trench, the central Pacific, Antarctica and elsewhere. In spite of the large number of fish examined, they recovered only the following adult metazoan parasites: 31 copepods (28 belonging to one species, 3 to another); about 600 digeneans (225 from one host species and 375 from another); 3 monogeneans apparently of 1 species; 3 cestodes. The following larvae were found. There were 221 fishes with larval nematodes, often with more than one worm in a host; 2 larval copepods; about 15 metacercariae; about 80 larval tetraphyllid cestodes; and 1 larval acanthocephalan (Noble, personal communication). The authors suggested that the Monogenea could have been washed off due to the extreme turbulence in the collecting device. However, monogeneans have repeatedly been recovered from deep sea fish, for example from the crossopterygian *Latimeria* and from macrourids (see below), and it seems unlikely that many monogeneans were lost in this way. Of 594 fish, 164 (28 per cent) had protozoan parasites, Myxozoa being the most common. Fishes from depths greater than 800 m had a greater frequency of infection than fishes from 100–800 m. The paucity of the deep water parasite fauna of pelagic fish is also shown by the data collected by Gusev (cited by Polyanski 1961a) who found only one skin copepod and no trematodes, cestodes, nematodes and acanthocephalans in 43 specimens of 5 bathypelagic fish species from a depth of 800–5,720 m in the Pacific, and by those collected by Orias *et al.* (1978), who examined 84 specimens of 5–8 bathypelagic fish species from a depth of 975–1,555 m in the eastern Atlantic. They recovered only 6 specimens of 1 trematode species from 1 host specimen and 1 specimen each of 2 copepod species from 2 other host specimens.

Bottom dwelling fish in the deep sea generally have a richer parasite fauna than the pelagic ones. Thus, the holocephalan fish, *Chimaera monstrosa*, which is a near bottom fish of the coastal north Atlantic from depths of 200–600 m, has a considerable number of ecto- and endoparasites. Dienske (1968) examined 215 fish from near Trondheimsfjord in Norway, and found the following species (adults if not stated otherwise): 2 Monogenea (1 from the gills, 1 from the cloaca), 1 copepod, 1 aspidogastrean trematode, 1 adult and 1 larval digenetic trematode, 3 cestodarians and 1 larval nematode. Other authors found in addition 1 digenean, 1 leech, 2 copepods and 1 isopod species (cited by Noble 1973). Protozoan parasites of this species have not been well studied.

Species of the fish family Macrouridae have been studied most extensively. The family comprises more than 300 species, most of which live close to the continental slopes at a depth of 200–2,000 m (Noble 1973). Most species are carnivorous bottom feeders and live up to 10 years, reaching a length of about 20–100 cm (Noble 1973). In most habitats, fish of such size, age and feeding habits could be expected to have a rich parasite fauna. Gusev (cited by Polyanski 1961a) recovered 2 species of copepod, 1 monogenean, 1 trematode, 2 cestodes and 2 nematodes from *Macrurus acrolepis* in the Pacific, caught at a depth of 2,500–7,000 m. Noble (1973) examined 275 fish of 17 species of this family from California, Newfoundland and the northern Atlantic. Although infections and diversity were more common than in the midwater fish discussed above, they were nevertheless much reduced in comparison with surface fish. Most common were Myxozoa and larval nematodes, but there were also Digenea, Cestoda, Copepoda, Monogenea and Microspora. The paucity of the parasite fauna is shown by the fact that only approximately 7 (?) species of copepods and probably 7 (or 6 ?) species of Monogenea were found (see also Armstrong 1975).

A greater diversity and higher infection rates were found by Campbell *et al.* (1980), who examined 1,712 fish of 52 benthic species including Macrouridae from depths of 53–5,000 m off the New York Bight (39–49°N, 70–72°W). They recovered a total of 17,007 helminths, comprising 80 identified species, and 137 copepods comprising 3 identified species. The authors concluded that although the deep sea fauna contains fewer families and genera than shallow water faunas, the deep sea parasite fauna is not unusual in terms of its abundance, diversity or host specificity. However, at the greatest depths, parasite abundance and diversity dramatically decline.

With Noble (1973), we can summarize that midwater fishes examined have far fewer species and numbers of parasites than inshore and open ocean surface fishes. There are very few adult helminths. In order of abundance the parasites are larval nematodes, myxozoans, larval cestodes, copepods and larval trematodes of the family Hemiuridae. The parasite fauna becomes poorer with depth until the benthopelagic zone is reached. Bottom dwelling fishes have more species and greater numbers of parasites than midwater species, because their food probably includes more infected intermediate hosts, and also because they are larger and have a longer life span. However, more studies are needed to clarify whether these generalizations hold for all latitudes. The difference in diversity of parasitic species between surface and deep water pelagic fish apparently exists only in regions where the surface waters are warm. It would be of special interest to compare deep sea parasite faunas in tropical and cold–temperate zones on the basis of surveys as large as that by Campbell *et al.* (1980).

Relict parasite faunas in isolated seas and hosts

A relict population or species is one which has been isolated from its parent group for a long time. It often lives in an isolated area due to contraction of an originally widely distributed fauna. For example, most fish of the family Gadidae are marine, but one species, *Lota lota*, is common in European and North American lakes and rivers, with no contact with the sea. Some of its parasites are otherwise found only in marine fishes, that is they represent a marine relict fauna of parasites in a freshwater fish (Zschokke 1933). Russian authors have made careful analyses of several such relict parasite faunas, and these were reviewed by Dogiel (1964), to whom the reader is referred for details. Dogiel concluded that relict parasite faunas are generally poor, and that relict animals often acquire some new parasites, characteristic not of its origin but of its habitat. One example discussed by him is the Caspian seal, *Phoca caspica*. It has 6 parasite species, of which 3 are also found in northern seals, whereas 3 are newly acquired, probably from birds. Another example from Dogiel is the parasite fauna of fishes in Lake Baykal. Approximately 20 of its fish species are of marine origin. Two of their parasite species are very ancient marine forms, and 8–9 species are typical freshwater parasites, derived from parasites of Cyprinidae and other freshwater fishes but representing distinct species.

Parasite endemicity of remote oceanic islands

Briggs (1966) compared the rate of endemicity of remote oceanic islands (300 miles or more from the nearest mainland) and found that it is very low in the north and middle Atlantic, markedly greater in the south Atlantic and Pacific, and exceedingly high in sub-Antarctic waters (Table 9). He suggested that, since marine animals are highly sensitive to temperature change, Pleistocene temperature reductions eliminated the endemic species of oceanic islands in some areas. According to this author, the major temperature drop took place in the northern and middle Atlantic, the change in the south Atlantic and Pacific was relatively small, and it was extremely small in sub-Antarctic seas. However, no Paleo-temperature determinations have yet been made to corroborate his suggestions, which are entirely based on faunistic findings.

Manter (1967), stimulated by the finding of Briggs, compared the endemicity of trematodes of remote oceanic islands. His data, supplemented by some data of Monogenea, are contained in Table 9. The greater endemicity of parasites is almost certainly due to the fact that they are less well-known than fishes. But the data suggest that the degree of endemicity of trematodes and monogeneans at various remote islands corresponds to that of their hosts.

Importance of temperature for parasite distribution

Thus far in this chapter it has been shown that there is a richer parasite fauna in the Pacific than in the Atlantic, a largely different composition of the parasite faunas in the two oceans, a greater endemicity of parasite taxa in the Pacific, gradients of increasing species diversity from cold to warm waters, and decreased diversity in

Table 9. Endemicity of marine fishes and their parasites of oceanic islands. From Briggs (1966; fishes), Manter (1967; Trematoda), Yamaguti (1968; Hawaiian Monogenea), Meserve (1938; Galapagos Islands Monogenea).

	Fishes	Endemicity of: Trematodes	Monogenea
North Atlantic			
Azores	0%	?	?
Madeira	3%	?	?
Bermuda	5%	2% (1 of 43 species)	?
Cape Verde	4%	?	?
South Atlantic			
St. Helena	27–50%	?	?
Tristan da Cunha	23%	?	?
Pacific			
Galapagos Islands	15–27%	61% (of 80 species)	(6 of 11 species)
Hawaii	34%	74% (of 92 species)	88% (128 of 146 species)
Easter Island	29%	?	?
Lord Howe Island	22%	?	?
Fiji	?	56% (of 35 species)	?
New Caledonia	?	45% (of 33 species)	?

the deep sea. I shall now try to determine which factors are responsible for these differences.

It is generally accepted that temperature is the most important single factor determining the distribution of marine organisms (Hedgpeth 1957). Temperature affects parasite faunas in two ways. Firstly, species numbers are greatly increased in warm seas and, secondly, species are different in cold and warm seas. That temperature is the decisive factor responsible for differences in species numbers is indicated by the observation that habitats at different latitudes which do not differ significantly in any other parameters beside temperature nevertheless differ in species numbers. Such habitats are the body surface and the gills of marine fish (Rohde 1978e), and it was shown above that species numbers of Monogenea per tropical fish species are greater than those for cold–temperate fish species. Rohde (1978d) postulated that higher temperatures accelerate the speed of evolution, probably by shortening generation times and increasing the speed of selection (and mutation rates?). Greater evolutionary speed must lead to the accumulation of more species in tropical seas and warm surface waters. None of the other explanations given by various authors for the diversity gradients is satisfactory (review by Rohde 1978d).

That temperature is a major factor determining the kind of parasite species in certain seas is indicated by a critical look at some surveys. Of the 98 parasite species of fish found by Shulman and Shulman-Albova (1953) in the White Sea, only 2 are cosmopolitan, whereas all other marine species (except for 37 with undetermined distribution) are known only from other cold and cold–temperate seas. According to the findings of Zhukov (1960), of the 87 species of endoparasitic helminths from the northwestern part of the Sea of Japan, many occur also in other cold seas of the Atlantic and Pacific regions (see above), but only 2 in the warmer southwestern part of North America and 2 in the southeastern part of North America. Of 87 species known in the cold northeastern and 92 species in the warm southeastern part of the Sea of Japan, only 21 species are common to both regions, whereas 36 species are common to the cool northwestern part of the Sea of Japan and the cool South Kuril Islands. The cold South Kuril Islands and the warm southeastern part of the Sea of Japan have only 11 species in common. Of 53 species recorded from the South Kurils, about half are also known from the Barents and White Seas, but only 2 and 1 from the warmer southwestern and southeastern parts of North America, respectively. The distribution of parasites corresponds to that of free-living organisms. The northwestern part of the Sea of Japan generally has cold water with free-living cold water organisms, whereas the southeastern part has warm water with a free-living subtropical fauna.

Importance of temperature for parasite faunas is also indicated by the distribution of some cosmopolitan species. One trematode, very widespread and not host-specific, is *Derogenes varicus* which has been found in more than 70 species of fishes in cold and temperate waters of the Pacific and Atlantic, extending from the Arctic to the Antarctic. At low latitudes it occurs only in deep cold water or in regions with cold currents. Thus, Manter (1955) found it in 5 different host species at Tortugas, Florida, all of which were from the cool water of not less than 250 m in depth. He once found the species also at the Galapagos Islands, where the water is cool and upwelling, because of the Humboldt current. These findings agree with the generalization that organisms of polar surface regions occur at greater depth at lower latitudes (Lamouroux 1816, cited by Hedgpeth 1957). Numerous examples for such "depressions" not only of particular organisms but also of entire communities are known (see for instance Hedgpeth 1957). Warm water species may conversely be "depressed" at higher latitudes, below the cold surface water (Hedgpeth 1957).

Finally, evidence for the importance of temperature is given by the comparison of shallow and deep water parasites in a particular region. This was done by Manter (1955) at Tortugas, Florida. Of 49 trematode species collected from fishes in deep water (100–1,000 m), only 5 were also found in the nearby surface water. Many deep water species at Tortugas indicate distinct cold water affinities. Of the trematode genera *Genolinea* (7 species), *Tubulovesicula* (6 species), *Genocerca* (3 species), *Hemipera* (4 species) and *Derogenes* (11 species), all are rather common in the north Atlantic and north Pacific, in deeper but not shallow water of the warm seas, and in the New Zealand and Tasmanian regions.

Importance of age of oceans for parasite diversity

Temperature may explain latitudinal differences in species diversity and composition of parasites but it cannot explain the differences in species diversity between the Pacific and Atlantic. We have to look for another explanation, and the one usually given is the difference in area of the two oceans. Probability of species formation in a large area is greater than in an otherwise identical smaller area, because chances of random isolations of populations are greater and such isolations are essential for geographical speciation. However, another explanation is at least as plausible: the older age of the Pacific Ocean. According to the now generally accepted Theory of Continental Drift, the Pacific has been in existence for much of geological time, whereas the Atlantic began to form only approximately 150 million years ago, when the continents began to drift apart. The assumption is reasonable that more species are found in the Pacific because there has been more time for species to originate and accumulate. A precondition for such an assumption is that evolution in the two oceans has been largely separate, and this is indeed supported by facts. It was noted above that there has been a considerable exchange of parasitic species between the cold northern Pacific and Atlantic, probably during the warm interglacials, and there also has been some exchange between the tropical eastern Pacific and western Atlantic across a narrow transoceanic connection which was closed several million years ago, and between the Indian Ocean and eastern Atlantic via the Tethys Sea existing until the Tertiary (Ekman 1953). Nevertheless, the great endemicity of genera and species in both ocean systems indicates that exchange has been limited. Furthermore, temperature differences at different latitudes would prevent spreading of introduced cold water forms into warm waters and of warm water forms into cold waters.

The difference in the age of the oceans can give a better explanation than larger area of greater *relative* species diversity of Monogenea (number of parasite species per host species) in the Pacific Ocean. The surface areas of fish (skin and gills) from different seas are identical, but species numbers nevertheless differ. The younger age of the Atlantic also explains reduced endemicity there. A young ocean would acquire a large proportion of its species from an old one, simply because the stock of its own species is relatively small.

Effects of host migrations on parasites; parasites as biological markers

Many species of marine animals migrate over long distances, and changes in the composition of their parasite faunas may occur during such migrations. The changes are considerable if migrations are between freshwater and the sea. The salmon, *Salmo salar*, has been particularly well examined. According to Dogiel (1964), the fish remain for three to four years in the northern Russian rivers, where they grow slowly. They then migrate to the sea and undergo rapid growth. After two or three years they return to the rivers for their first spawning, then after several months to almost a year they return to the sea for another one to two years, spawn again in the rivers, etc. Young salmon 1–4 years old gradually build up a freshwater parasite fauna consisting of 12 species. Adult salmon, after return from the sea, have 15 parasite species of which only three are freshwater species and these have been newly acquired. During the upstream migration, at first the marine ectoparasites and subsequently the marine intestinal parasites are lost. Parasites in the body cavity are least affected. Only few freshwater parasites are acquired because salmon do not feed during the upstream migration. The marine parasite fauna is reestablished after each return to the sea. Salmon in the River Exe, England, undergo similar changes (Kennedy 1975).

Changes in the parasite fauna may be small or non-existent if host populations migrate between different seas, and this facilitates use of parasites as biological markers.

Parasites repeatedly have been used to study populations and movements of marine fish. According to Sindermann (1957), there are some complicating and limiting factors of such studies: (1) kind and numbers of parasites may vary seasonally, geographically and with the age of the host, and parasite abundance may be affected by differential mortality, which may vary with season or host age; (2) parasite distributions may reflect variations in the distribution or abundance of intermediate or alternate hosts rather than the host being studied; (3) temperature and other physical factors may affect the parasite fauna; (4) it is possible that long term fluctuations in parasite abundance exist which are not revealed by short term studies. But Sindermann concludes that "despite these limitations, studies of the nature and degree of parasitization, when combined with other criteria for population analysis and with other methods" of studying fish movements, may give insights into the structure and movements of populations. Sindermann's (1957) study of parasites of North American herring, for instance, suggested that there is no mixing of adult herring of the Gulf of Maine and those of the Gulf of Saint Lawrence, and that immature herring from the eastern and western parts of the Gulf of Maine may be separate populations.

Some other examples of studies using parasites as biological markers are the attempt by Olsen and Pratt (1973) to use parasites for tracing nursery grounds of sole, *Parophrys vetulus*, the study by Grabda (1974) who found different populations of herring, *Clupea harengus*, in the Baltic Sea with different *Anisakis* infections, and that by Platt (1975), according to whom populations of cod, *Gadus morhua*, in the north Atlantic and Arctic, have different burdens of *Anisakis* and *Phocanema*. Hislop and MacKenzie (1976) used parasites in conjunction with tagging, to examine the question of whether they are more than one stock of whiting, *Merlangius merlangus*, in the northern North Sea. Hare and Burt (1976) could distinguish populations of Atlantic salmon, *Salmo salar*, from various tributaries of the Miramichi River system in Canada on the basis of their parasite fauna, and Platt (1976) found that populations of north Atlantic cod from Greenland and Iceland differed in their infection with larvae of the nematode, *Phocanema decipiens*. The author suggested that the relative abundance of the parasite could aid in determining the ratio of Iceland and Greenland components of the cod population on the spawning grounds of southwestern Iceland. Migratory and non-migratory freshwater populations of Arctic char, *Salvelinus alpinus*, could be distinguished by their different parasite faunas (Dick and Belosevic 1978), and MacKenzie (1978) showed that parasites of herring, *Clupea harengus*, permit tracing of its recruitment migrations. Capelin, *Mallotus villosus*, are infected with the cestode, *Eubothrium parvum*, in the Barents Sea and the Balsfjord, north Norway, but infections are heavier and overdispersed in the latter locality. Failure to find heavily infected fish in the Barents Sea, where the infection is underdispersed, was interpreted by Kennedy (1979) as confirming the suggestion that capelin in Balsfjord forms a local isolated population which does not migrate into the Barents Sea. For further examples see Gibson (1972) and Grabda (1976).

CHAPTER 9

Economic and Hygienic Importance of Marine Parasites

Parasitic diseases of marine fishes

Important monographs dealing with parasitic disease of marine fish are those by Schäperclaus (1954), Dogiel *et al.* (1961), Sindermann (1966, 1970, 1978), Williams (1967), Kabata (1970), Lom (1970), Mann (1970), Margolis (1970), Reichenbach-Klinke and Landolt (1973) and Grabda (1977b). A bibliography of books and symposia on fish parasites was given by Mann (1978). All fish species which have been examined to any degree are hosts to many species of parasites (for examples see Chapters 2 and 3). More than 550 species of Protozoa have been described from marine fish (Lom 1970), and there are more than 1,000 known species of Copepoda from fish (Kabata 1970) and thousands of species of trematodes and Monogenea. Only some examples of parasites important to fisheries can be discussed here.

The importance of fish parasites may be of several kinds: they may reduce population numbers by causing mass mortalities; they may affect the reproductive organs and reduce the number of offspring and thus population size; they may reduce the weight of fish; and they may make it more difficult and expensive to market fish.

There are several examples of mass mortalities caused by parasites, although evidence is often indirect. When the monogenean *Nitzschia sturionis* was transferred with its host, the sturgeon *Acipenser stellatus*, into the Aral Sea, an epizootic broke out among local sturgeon, *A. nudiventris*, which was not infected with the monogenean before. The population was so much reduced that it had to be placed under protection (Petrushevski and Shulman 1961). Petrushevski and Shulman (1961) also described a severe epizootic caused by the myxozoan *Myxobolus exiguus* in *Mugil cephalus*, the sea mullet, in the Black and Azov Seas. The gill filaments were strongly affected, and some fish were bleeding strongly from the gills.

In some parts of the "Tasmanskiy Zaliv" up to 500–600 dead fish per kilometre of shore line were observed.

The myxozoan *Kudoa clupeidae* infects only young Atlantic herring, *Clupea harengus*, and 75 per cent of year-old fish were found to be infected in certain areas (Sindermann 1966). According to Sindermann (1963), the infection is acquired in the early months of life and the symptoms disappear by the end of the third year. No differential mortality of infected herring could be observed in laboratory experiments and the percentage of infection remained unchanged between the first and second years of life. Nevertheless, it cannot be ruled out that in nature old fish are not infected because only parasite free fish survive. Sindermann (1966) gives as another example of mass mortality caused by parasites infection of smelt, *Osmerus eperlanus*, by the microsporan *Glugea*. According to Petrushevski and Shulman (1961), the combined effect of *G. anomala* and Monogenea led to mass mortalities of stickleback in the White Sea.

Many parasites affect the reproductive organs of fish and it seems possible that they may reduce the number of offspring and the population size of fish. The coccidian, *Eimeria sardinae*, parasitizes the testes of clupeoid fish like sardines, sprat and Atlantic herring. It infects more than half of the sardines in certain areas and often leads to sterility. Several other coccidians also infect the testes of clupeoids (Sindermann 1966). The microsporan, *Glugea hertwigi*, prevents reproduction by mechanical occlusion of the vent and it may also destroy gonad tissue in smelt, *Osmerus eperlanus*. The copepod, *Peroderma cylindricum*, retards development of the gonads of the European sardine and causes parasitic castration in a large proportion of the infected fish (Kabata 1970).

Many examples for the third case, reduction of weight, are known. Kabata (1970) quoted a study in Germany by Mann, who examined the effects of the copepod, *Lernaeocera*, on commercially important Gadidae. In three species, weight of fish infected with 1 or 2 parasites was reduced as follows: *Merlangius merlangus* (gutted) 0–20 per cent (1 parasite), 19–42 per cent (2 parasites); *Gadus morhua* (gutted) 0–28 per cent (1 parasite), 30 per cent (2 parasites), *G. morhua* (not gutted) 0–3 per cent (1 parasite), 0–35 per cent (2 parasites); *Melanogrammus aeglefinus* (gutted) 0–47 per cent (1 parasite); *M. aeglefinus* (not gutted) 0–40 per cent (1 parasite), 28–36 per cent (2 parasites). Corresponding to the weight loss is a retardation of growth due to the infection, although the effects are less marked and older fish are less strongly affected. Myxozoa in the bladder may lead to emaciation of the host and, according to Petrushevski and Shulman (1961), larvae of the nematode *Thynnascaris aduncum* in cod cause weight loss and reduced oil content and can in extreme cases completely destroy the liver (Fig. 71).

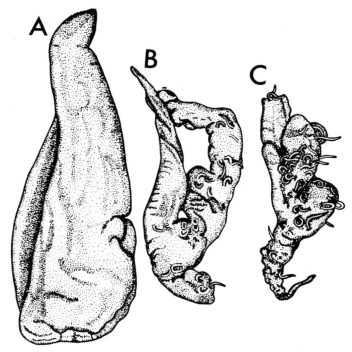

Fig. 71. Liver from three specimens of Baltic cod of equal size. A – healthy liver; B and C – infected with nematode *Thynnascaris aduncum* (from Petrushevski and Shulman 1961, after Shulman).

The fourth point, increased difficulty and expense in marketing fish because of parasite infection, is exemplified by some myxozoans, nematodes and trematodes. Among the Myxozoa, *Unicapsula*, *Chloromyxum* and *Kudoa* infect many fish such as mackerel, halibut, herring, stockfish, John Dory, etc. In Australia, the barracouta or snoek, *Thyrsites atun*, may harbour *Kudoa*, which produces "milkiness" of the fish (Johnston and Cleland 1910, Willis 1949, Roughley 1951). Five per cent of the fish may be infected. Rohde (1976c) reported that all large kingfish, *Seriola grandis*, at the southern end of the Great Barrier Reef were infected with *Kudoa*, which makes the flesh tasteless and inedible. Lester (personal communication) found *Unicapsula* in kingfish near Brisbane, Australia. Fish infected with myxozoans in the muscles show rapid deterioration of the flesh after death and therefore are much more difficult to market. In U.S.A., the microsporan *Pleistophora macrozoarcidis* is responsible for tumour-like cysts in the muscles, several centimetres in diameter, which reduce the marketability of the fish (Sindermann 1966), and the mullets, *Mugil capito* and *M. cephalus*, cultured in Israel, were found to be infected with metacercariae of *Heterophyes* which reduce the quality of the flesh (Lahav 1974).

Nematode larvae mainly reduce the value of fish because of the unappealing appearance they give to the flesh. Infections of various marine fish such as cod, smelt, plaice, etc. with the "codworm", *Phocanema decipiens*, may result in reduced value, and coral trout, *Plectropomus leopardus*, and other valuable reef fish in Australia, are often heavily infected with black nematode larvae in the mesenteric tissue and viscera (personal observation).

Parasitic diseases of marine molluscs

Korringa (1952b) gave a review of diseases and parasites of oysters, with an extensive bibliography, and Cheng (1967), Sindermann and Rosenfield (1967) and Sindermann (1970) discussed parasites of molluscs in general. Bibliographies of diseases of commercial bivalves were given by Sindermann (1968, cited by Key and Alderman 1974) and Key and Alderman (1974); Quayle's (1975) bibliography on tropical oyster culture contains some references to helminths of oysters, and Comps (1978) reviewed advances in oyster pathology including parasitic diseases. Many of the parasites do not damage their hosts or do so only to an insignificant degree, but some may result in great losses to fisheries.

Oyster parasites Oyster beds have been repeatedly devastated but because of insufficient knowledge in the past the causes of early destructions are not known. According to Korringa (1952b), devastating mortalities of the European oyster, *Ostrea edulis*, occurred in 1920–21 and many oyster beds never recovered. The cause is unknown but disease was probably involved, although abnormal weather conditions have been blamed. Mass mortalities of the American oyster, *Crassostrea virginica*, in eastern Canada occurred in 1915 and in the following years all important oyster beds in Prince Edward Island were destroyed. Again, Korringa suggested contagious disease as the cause. Repopulation occurred by new resistant stocks.

With accumulating knowledge, agents responsible for later mortalities could be determined. According to Sindermann and Rosenfield (1967), a drastic decline of oyster production in Delaware Bay, U.S.A., began in 1957, in Lower Chesapeake Bay in 1959. Mortalities often exceeded 95 per cent for several years. The ascetosporan, *Minchinia nelsoni*, was found to be responsible (Fig. 16; for details of structure and parts of the life cycle see Haskin *et al.* 1966, Farley 1967, and Perkins 1968). Other ascetosporans infecting oysters are *Minchinia costalis*, which may co-occur with *M. nelsoni* in the same oyster individual (Couch 1967), and *Chytridiopsis ovicola* which infects eggs of European oysters (Léger and Hollande 1917). The gregarine *Nematopsis ostrearum*, whose second host is crabs, has been held responsible for extensive oyster mortalities in Virginia and Louisiana, although it may not be the direct cause (Sindermann and Rosenfield 1967), because it has been observed that there is not always a correlation between infection and mortality (review by Korringa 1952b). A common gut parasite of oysters is the flagellate, *Hexamita* (Schlicht and Mackin 1968). It has been thought to be responsible for mass mortalities of oysters in Holland ("pit disease"), but its role is uncertain. Scheltema (1962) found *Hexamita* in a large percentage of oysters in Delaware Bay, but only during winter and early spring. Maximum infection intensity coincided with the minimum water temperature, and temperatures over 25°C were lethal to the parasite. It is, thus, possible that *Hexamita* acts as a pathogen only at low temperatures.

Several ciliates infect oysters, such as *Sphenophrya* which forms large cysts on the gills. Amoebas like *Vahlkampfia* have also been found, but there is no evidence for pathogenicity (Hogue 1921). Larval trematodes of the genus *Bucephalus* can cause parasitic castration of oysters but "infected oysters have an excellent flavor and are fat-looking and glycogen-rich throughout the year, whereas normal oysters are thin and relatively tasteless during part of the year" (Hopkins 1957, cited by Sindermann and Rosenfield 1967). In New Zealand, larval Bucephalidae led to oyster mortality under experimental conditions (Millar 1963) and other trematode larvae may halt growth and inhibit reproduction.

Among the cnidarians, turbellarians, annelids and mollucs are some species which may be either commensals, parasites or predators of oysters (see Chapter 2 and Pearse and Wharton 1938, Korringa 1952b, 1954). The small pyramidellid snail, *Menestho bisuturalis*, has its feeding position along the edges of oysters, to which the snails attach themselves by their oral sucker; the proboscis is inserted into the soft tissues and they interfere with the normal development and growth of their hosts, as indicated by the characteristically deformed shells of young oysters from heavily infected areas (Losanoff 1956). Another snail, the slipper limpet, *Crepidula fornicata*, is a predator rather than a parasite of oysters. Among the sponges, *Cliona* bores into the shells of oysters and causes mortality (Korringa 1952b).

Some parasites of oysters belong to the Crustacea. The crab *Pinnotheres ostreum* (see Christensen and McDermott 1957, and Christensen 1958 for life history and biology) lives in *Crassostrea virginica*, injuring its gills. It has been claimed that infection leads to mortality (Korringa 1952b, Sindermann and Rosenfield 1967). The copepod, *Mytilicola orientalis*, infects oysters in Japan and has been transferred with seed oysters to the United States. It may have pathological effects. The related species *M. intestinalis* may lead to widespread mortalities in mussels, although this is questionable (Bayne et al. 1978). It infects oysters only rarely. A key to the species of *Mytilicola* was given by Humes (1954).

In Australia, Haswell (1886) described the polychaetes *Polydora ciliata* and *P. polybranchia* from the Hunter River in Australia, the first species supposedly identical with the European species, and thought them to have caused the death of large numbers of oysters. Wolf (1972) suggested that high mortality of Sydney rock oysters, *Crassostrea commercialis*, in Moreton Bay near Brisbane, might be due to infection by an ascetosporan, later described as *Marteilia sydneyi* by Perkins and Wolf (1976). The pathogen occurs in tropical and subtropical regions of eastern Australia, but not in the colder waters south of Richmond River. The blacklipped oyster, *Crassostrea echinata*, from eastern Australia is probably parasitized by the same species (Perkins and Wolf 1976). Other parasites known from Australian oysters are larvae of the cestode *Tylocephalum* in *Crassostrea commercialis* from northern New South Wales and southern Queensland and in *C. echinata* in the Northern Territory (Wolf 1976b), an unidentified unicellular organism in the eggs of *C. echinata* in northern Australia (Wolf 1977) and an unidentified unicellular organism in the digestive diverticula of the pearl oyster, *Pinctada maxima*, in tropical Australia (Wolf and Sprague 1978). Spores of the latter organisms were found only in moribund oysters and the infection may be responsible for mass mortalities which have affected some commercial pearl farms. *Tylocephalum*, although usually not fatal, greatly reduces the condition of infected oysters (Wolf 1976b).

Mussel Parasites Various protozoans, helminths and crustaceans infect mussels. Examples of parasites which are pathogenic are the ascetosporan, *Haplosporidium tumefacientis*, which is responsible for swellings of the digestive gland and kidney of the California mussel, *Mytilus californianus* (Taylor 1966), thigmotrichid ciliates which infect *Mylitus edulis* in California (Pauley *et al* 1966), the trematode, *Proctoeces maculatus*, in the same host (Tripp and Turner 1978) and the coccidian, *Pseudoklossia pectinis*, in the kidney of scallops. However, the copepod *Mytilicola* is probably most important to the mussel industry, although even this is not beyond doubt. According to the discussion by Sindermann and Rosenfield (1967), *Mytilicola porrecta* in the Gulf of Mexico does not cause mortality or pathological symptoms, and there is no correlation between infection with *Mytilicola* and condition of mussels on the Yugoslav Adriatic coast. Apparently, the density of mussel populations is important. Nevertheless, *Mytilicola* appears to be a severe threat to the mussel industry in Europe, but it is not clear whether the parasite is the direct or only an indirect cause (see page 167). Davey and Gee (1976) suggested that beside population density and location of the hosts, degree of exposure to wave action is important in determining the infestation levels. The parasite burden also increases with host size, which is unlikely to represent an accumulation of parasites with increasing host age because of the short life span of *Mytilicola*. The authors consider it possible that higher filtration rates of large hosts are responsible. Perkins *et al.* (1975) described the ascetosporan, *Urosporidium spisuli*, which infects an immature nematode, probably *Sulcascaris sulcata*, parasitic in surf clams, *Spisula solidissima*, on the North American Atlantic coast (Sprent 1977, Lichtenfels *et al*, 1978). When the hyperparasite sporulates, its spores give the worm a brownish-black appearance which results in clams being withheld from sale. Scallops, *Amussium balloti*, in Queensland, Australia were found to be infected with larval nematodes by Cannon (1978a). Twenty five of 467 scallops were infected and the parasites were surrounded by brown or orange tissue which may reduce the marketing value. Infection of the scallop, *Pecten alba*, with a trematode was shown to lead to parasitic castration (Sanders 1966).

Parasitic diseases of marine crustaceans

Sindermann and Rosenfield (1967) gave a review of diseases in commercially important crustaceans, with an extensive bibliography, and a chapter dealing with parasite infections of marine crustaceans can also be found in Sindermann (1970). Crabs, lobsters and shrimp have many species of Protozoa. The ascetosporan *Minchinia louisiana*, related to species responsible for mass mortalities in oysters, infects crabs on the North American Atlantic coast, and *Urosporidium crescens* infects metacercariae of a trematode in crabs, *Callinectes sapidus*. The metacercariae become dark as a result of infection with the hyperparasite and this "pepper crab disease" prevents sale of the crabs (Perkins 1971). The microsporan *Nosema* may be of some economic importance to crab fisheries on the North American Atlantic coast and it affects shrimp populations in the Gulf of Mexico significantly. Species of the microsporan *Thelohania* are common parasites of shrimp in Europe, North and South America, and gregarines of the genus *Nematopsis* and others infect crabs, lobsters and shrimp. Various ciliate genera are found in and on crabs and shrimps, and one species, the suctorian, *Ephelota gemmipara*, infects and destroys eggs on female Norwegian lobsters, leading to a 90 per cent decrease in production of larvae in hatching boxes. Many species of larval trematodes (sometimes with protozoan hyperparasites), larval cestodes, some acanthocephalans, turbellarians, nemerteans, nematodes and leeches are also known from crabs, lobsters and shrimp but most are economically unimportant. With regard to parasitic crustaceans, crabs are affected by isopods, copepods and Rhizocephala. Infection with the last, of which *Sacculina* is an important genus, leads to the destruction of the host's gonads resulting in sterilization and may thus perhaps affect host populations (see page 158). Several species of isopods and rhizocephalans are also known from shrimps and the blood sucking copepod, *Nicothoe astaci*, infects the gills of European lobsters.

Parasitic diseases of marine mammals

Sweeney (1974) reviewed diseases of pinnipeds, and Sweeney and Ridgway (1975) those of small cetaceans. Checklists of parasites of marine mammals in general and of Australian marine mammals were given by Dailey and Brownell (1972) and Arundel (1978), respectively. The latter, whose report is based on 57 references, also gives details of life cycles and diseases. Chapters in books dealing with parasites are those by Yablokov *et al.* (1972) who discuss helminths of whales and dolphins and their geographical distribution, and by Ridgway *et al.* (1975) who give some details of helminths of pinnipeds.

Many parasitic worms have been recovered from marine mammals kept in captivity, and frequently severe pathological symptoms associated with the infections have been observed. Some examples of parasitic disease in marine mammals that died in a marine zoological park were described by Cordes and O'Hara (1979). The nematode *Anisakis simplex* was found to have caused gastritis and ulceration in a sea elephant, cestodes of the genus *Diphyllobothrium* had obstructed the intestine of a sea lion, and the trematode *Campula palliata* caused parasitic cholangiohepatitis in 4 and possibly 6 of 12 dolphins. Three California sea lions which died in two aquaria in Japan had lungworm, probably *Parafilaroides decorus*, and one had a verminous pneumonia (Ashizawa *et al.* 1978). A female narwhal, *Monodon monoceros*, died of a lung infection with the nematode *Halocercus monoceris* (see McNeill *et al.* 1975).

That parasite infections resulting in disease are not restricted to animals kept in captivity was shown by Bishop (1979), who found natural infection with trematodes (family Campulidae) in a bearded seal, *Erignathus barbatus*. The infection was probably responsible for significant lesions in the pancreas and liver. Vauk (1973) reported the results of the examination of 12 seals, *Phoca vitulina*, found dead or dying in the North Sea. Although the most common cause of death appeared to be oil pollution, infection with the nematode *Otostrongylus circumlitus* may also have been important. Trematodes, such as *Opisthotrema*, may infect the nasal passages of dugong and cause severe pathological changes (Budiarso *et al.* 1979).

A common phenomenon is the beaching or stranding of whales, dolphins and other marine mammals. A variety of explanations has been given but little information on parasitic infection and its role in stranding is available (Dailey and Walker 1978). Several observations indicate that parasites could be an important factor. Stroud and Roffe (1979) necropsied 68 marine mammals of 10 species which had stranded on beaches in Oregon, U.S.A. They diagnosed heavy infections with endoparasites as the primary cause of death in 9 cases and as a contributory cause in another 9 cases. Seven young animals had worm induced pneumonia, and one *Eumetopias jubata* had died from anaemia due to damage of the stomach mucosa by nematodes. According to Geraci *et al.* (1978), 14 of 30 stranded female Atlantic white sided dolphins, *Lagenorhynchus acutus*, had *Crassicauda grampicola* in the mammary glands. The parasite leads to significant tissue reactions and may reduce milk output and hence herd productivity. Dailey and Walker (1978) and Stroud and Dailey (1978) observed that individual, single stranded cetaceans in southern California in the process of stranding exhibited a disoriented behaviour with an obvious loss of equilibrium. They recovered adult trematodes of the genus *Nasitrema* in the brain of 26 of the 40 stranded animals, with lesions most commonly on and in the cerebral hemispheres. The role of the infection in mass strandings is unresolved, but it seems at least possible that parasites are the cause or a contributing factor. According to the same authors, preliminary findings indicate that on the east coast of the United States, nematodes of the genus *Stenurus* located in the ears are a contributing factor to stranding. Nematodes were also found in the ear canals of beached whales by Hamrick (1974). Dailey and Stroud (1978) dissected 10 cetaceans of 5 species and found many species of helminths. They concluded that the parasites contributed to the stranding of 3 animals and led to death by secondary infections in a further 2. Hearing disturbances and changes in acoustic behaviour may in some way be involved in stranding, and Dailey and Ridgway (1976) discussed the possibility that parasites may be responsible for such disturbances and changes.

Economic importance of marine parasites

Kabata (1970) pointed out that quantitative assessments of losses due to marine parasites are rarely possible and that generally our chances of observing mass mortality in the open sea are negligible. Nevertheless, the foregoing discussion of parasitic diseases in marine animals has shown that parasites lead to considerable losses, although it is not always possible to pinpoint the actual cause of mortalities or to separate the share of marine parasites and of other factors in such mortalities.

Some estimates may illustrate the role of marine parasites. Kabata (1955, cited 1970) estimated that the effects of the copepod, *Lernaeocera branchialis*, on the haddock, *Melanogrammus aeglefinus*, would have caused a loss of approximately 100,000 kg of haddock to the Scottish fishing industry in 1955, if each copepod would cause a loss of 29 gram in each fish it infects. According to Kabata (1970), this is a decided underestimation. Approximately 10 per cent of Katfisch (catfish) were lost to human consumption in spring 1952 in Hamburg, Germany, because of infection with microsporans (Mann 1954). Hargis (1958) estimated that candling and trimming of fish to detect and remove parasites may increase the cost of packaging by up to 80 per cent.

The difficulty in exactly estimating the losses due to parasites (or commensals) is indicated by the case of the polychaete, *Polydora*. Approximately 30 per cent of oysters in South Carolina, U.S.A., were found to be infected (Lunz 1940, 1941), and the infection resulted in "considerable financial loss". Apparently, much of the loss is due to reduced quality of the oysters because of "mud blisters" which break during shucking, spreading mud and detritus all over the meat. Owen (1957) believed that death of oysters is not due to the worm as such, but that it contributes to the formation of a poor environment and in most cases is only indicative of such an environment. Similarly, spreading of the copepod of mussels, *Mytilicola intestinalis*, into many coastal regions of Europe in which the parasite was previously unknown led to an epidemic in the early 1950's which drastically reduced mussel production in parts of Europe, particularly in Holland (Meyer and Mann 1950, Korringa 1968). However, the parasite may not always be the only or even the most important factor in the decline. Thus, in the Wadden sea region of the Netherlands, mussel production declined from 44,500,000 kg in the 1954–55 season to 23,600,000 kg in 1955–56, but Korringa (1957) held the great average age and the poor quality of the mussels planted in early 1955 as mainly responsible. Nevertheless, *Mytilicola* appears to be the causative agent of extensive mussel mortalities (Korringa 1968), although some recent publications seem to indicate that it may be less pathogenic than often thought (Moore et al. 1978). More clearcut are oyster epidemics due to ascetosporans, some of which were discussed above. Oyster landings in New Jersey waters of Delaware Bay, U.S.A., in the late 1940's and early 1950's were about 6 million pounds of shucked meat annually until the mid 1950's. In 1960, they declined to 167,000 pounds due to infection with the ascetosporan, *Minchinia nelsoni*, and no significant recovery had occurred until the mid 1960's (Sindermann and Rosenfield 1967). The same species affected oysters in 1959 so strongly that commercial planting of oysters in parts of Virginia ceased in May, 1960 (Andrews and Wood 1967). The protozoan *Minchinia costalis* caused epidemics in oysters in Virginia. Losses ranged from 12–14 per cent in 1959 to 36–44 per cent in 1960 (Andrews et al. 1962).

In Australia, oyster cultures have been affected by mass mortalities, but the cause could not be ascertained. Roughley (1926) investigated mortalities of oysters on the George's River, New South Wales, which in eight or nine years up to 1924 had killed many oysters during the winters, in some years the majority of the marketable oysters. Many of the surviving oysters in winters of heavy mortality had abscesses, ulcerations or inflammation of various tissues. Although the cause of the pathological symptoms could not be determined, it was thought that this probably had a bacterial origin. Low temperatures were thought to reduce the resistance of the oysters to such infections. However, parasitic infections cannot be excluded.

Control of parasitic diseases of marine animals

For marine animals kept in aquaria, ponds or other isolated bodies of water, drug treatment, usually by adding chemicals to the water, may be useful. A list of drugs used in the control and treatment of fish diseases was given by Reichenbach-Klinke and Landolt (1973); but altogether, very little is known about drug control of marine parasites. Some of the methods used for the control of freshwater parasites, which are better known, may be adapted to the marine environment. Kabata (1970), for instance, discussed in detail various treatments used for crustaceans ectoparasitic on freshwater fish. Treatment of parasitic infections in marine fish cultures was discussed by Ghittino (1974), McVicar and Mackenzie (1977) and Sindermann (1977).

Many parasites are more sensitive to environmental factors than their hosts, and infection with such parasites is decreased in cultures. Möller (1976), for instance, kept *Gadus morhua* for 46 days, and *Zoarces viviparus*, *Myoxocephalus scorpius* and *Platichthys flesus* for 76 days in aquaria and found that infection intensities of five common intestinal parasites went down during this period, most strongly in those species which feed on the host's predigested food and are not, or only slightly, attached to the intestinal wall (*Thynnascaris aduncum* 93 per cent, *Echinorhynchus gadi* 85 per cent). Williams and Phelps (1976) reported that the number of parasite species of five species of marine fish decreased after cage culture. On the other hand, for certain parasites artificial conditions are optimal and they consequently increase enormously in numbers, leading to disease and death. Thus, Williams and Phelps (1976) found an increase in numbers of the ciliates *Scyphidia* and *Trichodina* on marine fish cultured in cages, and fish in marine aquaria commonly die of parasite infections.

Care should be taken to select the right food for cultured animals. For example, turtles bred in captivity may acquire heavy infections with nematodes when fed with fish harbouring infective larvae. Furthermore, a high density of populations kept in captivity enhances the probability of cross-infection and epidemics, and the optimal density for culturing marine animals should be chosen.

Diseases of marine animals in their natural environment are more difficult to control and only in a few cases have attempts at control been made. Sindermann (1970) lists the following points which should be considered:

1. Avoid the transfer of susceptible animals into epizootic areas, or of animals from epizootic areas;

2. Develop disease resistant stocks by selective breeding of survivors;
3. Accumulate basic information on the life history and ecology of the disease agent to define vulnerable stages or restrictive environmental requirements;
4. Maintain production in an artificial environment where the disease can be controlled.

The first point may be less important if intermediate hosts are necessary for the completion of the parasite's life cycle; transfer of infected animals into areas without these hosts may then be possible without danger of introducing the disease. An example related to the second point is the development of resistant strains of oysters after epizootics due to the protozoan *Minchinia nelsoni*. Existence of genetically determined resistance to parasite infection in invertebrates is also indicated by the observation that banded individuals of the Australian snail, *Velacumantus australis*, are less likely to be infected with larval trematodes than unbanded individuals (Ewers and Rose 1965). *Minchinia* can also be used as an example of point (3). It survives only in salinities higher than 15 per cent, and restriction of oyster planting during epizootics to low salinity areas or a temporary transfer can control the disease. Hatching methods for oysters and clams show that the method of point (4) is possible (Sindermann and Rosenfield 1967).

In addition to these general points, Sindermann gave special advice for inshore fishes, where disease can be controlled by deliberate overfishing, seeding of inshore areas with pathogens to maintain resistance of the population and prevention of returning viscera or entire diseased fish into the sea. Petrushevski and Shulman (1961), discussing an outbreak of *Myxobolus exiguus* in sea mullet, *Mugil cephalus*, in the Black and Azov Seas, also suggested that intensive fishing of the affected fish populations is the most effective control measure. It restricts the number of infected fish and permits utilization of fish before they become diseased.

Ivanov and Markov (1973) reported some preliminary findings according to which hybrids of *Acipenser ruthenus* and *A. sturio* are more resistant to infection; thus, the possibility of breeding resistant strains should be considered.

It may be important for parasite control of marine fish that parasites are often more sensitive to changes in external conditions than the hosts (see above). This was, for instance, shown for gill parasites of *Leptococcus armatus* and *Atherinops affinis* in Californian estuaries (Baker 1977). Populations from such extreme habitats may survive epidemics and can be used for restocking in decimated areas. How a good knowledge of the parasite's biology can lead to avoidance of losses is shown by the example of oysters infected with the protozoan, *Minchinia costalis*. Losses occurred mainly during a short season in May and June, and heavy losses could be avoided by careful timing of oyster planting and early harvesting (Andrews *et al.* 1962).

Based on a thorough study of host and parasite biology, various authors have recommended control measures for certain parasites.

For controlling *Mytilicola intestinalis* in mussels, Hepper (1955) recommended growing mussels on stakes, fences or ropes, in fast moving water and brackish water, because infections under these conditions are usually light. Korringa (1957) furthermore recommended that only half grown mussels be planted in *Mytilicola* infested areas.

For controlling *Hexamita* in oysters, Laird (1961) recommended elimination of overcrowding, use of shell to harden the muddy bottom, regular removal of self-silt and other deposits by dredging and other methods, removal of mud crabs and occasionally, depending on the local conditions, rotation of oyster crops and bottom ploughing. According to Laird, the use of different strains of oysters, of hybrid strains and of strains selected for certain conditions may also be valuable.

Human infections with marine parasites

Although the most common and dangerous parasites of man are associated with the terrestrial and freshwater environments, there are many marine parasites which are dangerous to man.

For a detailed account of marine helminths of man, the reader is referred to the review by Williams and Jones (1976) which is based on more than 300 references and also discusses pathological, preventive and historical aspects. Part of the following brief account is based on this review which lists the following marine cestodes, trematodes and nematodes occurring in man: Cestodes (*Diphyllobothrium latum, D. dendriticum, D. lanceolatum, D. ursi, D. dalliae, D. alascense, D. pacificum, Diplogonoporus balaenopterae, D. grandis, D. fukuokaensis, Spirometra*); trematodes (many species of Heterophyidae, including *Heterophyes heterophyes, Metagonimus yokogawai, Pygidiopsis summa, Haplorchis yokogawai, H. taichui, Stellantchasmus falcatus, Apophallus venustus, Nanophyetus schikobalowi,* and larval Schistosomatidae such as *Cercaria littorinalinae, Ornithobilharzia canaliculata, Austrobilharzia variglandis, A. terrigalensis*); nematodes (larval *Anisakis, Contracaecum, Phocanema, Trichinella*).

Of the species of *Diphyllobothrium* listed, the first six occur mainly or entirely in northern regions (northern Europe, Siberia and Alaska). *D. latum* is usually acquired by eating raw freshwater fish. *D. pacificum* is a common parasite of pinnipeds and infects man in Peru (Baer et al. 1967). *Diplogonoporus balaenopterae* and *D. grandis* sometimes infect man in Japan, and the third species, *D. fukuokaensis*, also has been reported from Japan. Normal hosts of *Diplogonoporus* are probably marine pinnipeds and cetaceans. All infections with these two genera are acquired by eating raw fish, and the symptoms, best known in *Diphyllobothrium latum* infection, may be severe, including a severe anaemia.

Numerous species of digenetic trematodes have been found in man. Most infections with adult worms are due to small flukes of the family Heterophyidae. They are quite non-specific for their hosts and most are acquired by eating raw freshwater, brackish or marine fish, but marine invertebrates such as shrimps may also be involved. In some parts of Japan, the Philippines and Polynesia such infections are very common (Laird 1961). *Nanophyetus salmincola* causes fatal salmon poisoning disease in dogs and other canines in North America. The intermediate host is a freshwater snail and the symptoms are due to the hyperparasitic microorganism *Neorickettsia helminthoeca*. No human infections with this parasite are known, but in some parts of Siberia almost whole human populations are infected with the related (or identical?) species, *Nanophyetus schikhobalowi*. That the infection is also associated with the marine environment is indicated by the occurrence of larval trematodes in ocean caught fish. Constipation or diarrhoea, abdominal pain and severe discomfort are the symptoms in human infection.

All human infections with marine schistosomes are due to the tailed larval stage, the cercaria, which attempts to penetrate into the skin. Sea birds are the normal hosts and the larvae do not succeed in migrating to the liver and blood system of man. Instead, they die in the skin causing an often severe dermatitis ("swimmer's itch", "clam digger's itch", "seabather's eruption", "weed itch", etc.). In Australia, a dermatitis caused by marine schistosome cercariae has been described by Bearup (1955, 1956). The cercaria of *Austrobilharzia terrigalensis* was shown to infect 4 per cent of the snail *Pyrazus australis* in Narrabeen Lagoon, N.S.W., where cases of a dermatitis after bathing have been reported. A dermatitis could be experimentally produced with the cercaria. At Heron Island, Great Barrier Reef, larvae of probably the same species infect the marine snail, *Planaxis sulcatus*, and adults parasitize the silver gull, *Larus novaehollandiae*, and the reef heron, *Egretta sacra* (Rohde 1977d).

Attempts to produce a dermatitis with the cercariae failed. At the same locality, *Gigantobilharzia* sp. was found in silver gulls (Rohde 1978c). Species of this genus cause a dermatitis in several countries. According to Ewers (1961), another species of bird schistosome develops in the false limpet, *Siphonaria denticulata*, and is responsible for a marine dermatitis in rock pools along the New South Wales coast.

Reichenbach-Klinke (1975), Margolis (1977), Smith and Wootten (1978), among others, discussed nematode larvae in marine fish and their importance to man. The infection of man by marine nematodes of the family Anisakidae is called anisakiasis. Species of three genera are involved, *Anisakis*, *Contracaecum* and *Phocanema*. Natural first hosts are marine crustaceans, and second intermediate hosts are marine fish and squid, which pass the infection on to the final host when eaten. Man becomes infected by eating raw or insufficiently cooked marine fish, squid and other invertebrates. The larvae penetrate through the intestine or stomach wall causing a gastrointestinal inflammation (Hayasaka *et al.* 1971, Shiraki 1974, Kagei *et al.* 1978, Yoshimura *et al.* 1979). The infection is (or was) not rare in countries where raw or pickled marine fish is commonly eaten, for example in the Netherlands, where people like to eat raw pickled "green" herring ("herring worm disease"), or in Japan, where people eat raw marine fish as "sushi" and "sushimi". Beside herring (Sluiters 1974, Van Banning and Becker 1978), many other fish species may serve as a source of infection. Stern *et al.* (1976), for instance, found 16 species of commercially caught fish infested with anisakid larvae, and Cannon (1977a,b) demonstrated larvae of *Anisakis, Terranova, Contracaecum* and *Thynnascaris* in 47 fish species in southeastern Queensland, Australia (see also Bussmann and Ehrich 1979, Grabda 1976, 1977a, 1978, Wootten 1978, Myers 1979, Payne *et al.* 1980).

The account given by Williams and Jones (1976) makes it clear that the probability of infection can be reduced by gutting fish at sea, which prevents migration of the larvae into the tissues of the fish, although some larvae are already in the tissues. Smith and Wootten (1975) showed experimentally that immediate gutting of herring reduces the number of *Anisakis* larvae in the flesh. Infection also can be prevented by heavy salting (slight salting is not safe), freezing to at least $-20\,°C$ for at least 24 hours, marinating for over 30 days with a high initial concentration of acid and salt, and smoking of salted (but not fresh) fish (Ruitenberg, cited by Williams and Jones 1976). Todorov (1973) found no live larvae after freezing for 2–3 months at -18 to $-20\,°C$, and Parukhin and Todorov (1977) reported that larvae were killed by freezing at -18 to $-26\,°C$ for 1 to 2 days (see also Dailey

1975, Hauck 1977). The safest way of preventing infection is, of course, to eat only cooked fish and invertebrates.

The effect of control measures against anisakiasis is shown by the fact that no new cases with the diseases were reported after the Dutch government introduced legislation to deep freeze fish used for raw consumption, in spite of the increase of infection in fish in the North Sea since the early 1950's (Rae 1972, Wootten and Waddell 1977).

Trichinella infections were common in many European countries, until strict meat inspections were introduced. Infection is usually acquired by eating raw or insufficiently cooked pork. Heavy infections may lead to death. The parasite inhabits a wide range of hosts and is found in many marine vertebrates, such as polar bears, seals, walrus and whales. Kozlov (1971) discussed ways in which pinnipeds become infected. Eskimos are frequently infected, probably by eating infected polar bears and walrus. Kozlov and Berezantsev (1968) demonstrated the parasite in walrus, and Thing *et al.* (1976) tabulated findings of *Trichinella* in this host. In an incident in 1975, 28 of 45 people acquired *Trichinella* infections in North America by eating walrus meat (Juranek and Schultz 1978; see also Thing *et al.* 1976). Even fish have been implicated in the cycle, but their role is not clear.

Adults of the nematode *Angiostrongylus cantonensis* naturally infect the brain and lungs of rats, which become infected by eating various invertebrates containing infective larvae. The parasite has been implicated as an agent of human "eosinophilic meningoencephalitis" in the Pacific region, and although freshwater invertebrates are the usual transmitters, experimental evidence indicates that marine bivalves and crustaceans may also be involved (Cheng 1966a).

The recently discovered nematode *Echinocephalus sinensis*, the larvae of which occur in a high percentage of oysters from Hong Kong (Ko *et al.* 1974, Ko 1975, Cheng 1976) is of potential danger to man. Successful experimental infection of a cat, kittens, monkeys and puppies indicates that man may be a host (Ko *et al.* 1975, Ko 1976, 1977). Deep frying or brief (30 seconds) dipping into boiling water killed the worms (Ko 1977). A related non-marine species (*Gnathostoma*) infects man in some countries, producing lesions and abscesses in the skin, subcutaneous tissue or muscles.

CHAPTER 10

Future Research

Taxonomy

Most studies of marine parasites have concentrated on cold and temperate seas in the northern hemisphere, and on parasites of vertebrates. Rohde (1976a, 1977b) estimated that the approximately 1,000 fish species at the southern end of the Great Barrier Reef are infected with at least 20,000 species of parasites, and very few of these are known. Thus, descriptive taxonomy, which is the basis of all ecological as well as physiological and morphological work, deserves the greatest attention, particularly in the tropics and the southern hemisphere.

Species formation

It had been believed for a long time that the only way of speciation is by means of geographical isolation (allopatric speciation). However, studies on insect parasites of plants have shown that one or a few mutations are sometimes sufficient to bring about reproductive isolation within a geographical area (Tauber and Tauber 1977a,b, Tauber *et al.* 1977). Sympatric speciation, or speciation without geographical isolation, is thus possible, but it is not known how common it is, and nothing about it is known in the marine environment. Price (1980) has pointed out that closely related and sympatric species are too numerous in some groups to be accounted for by geographical speciation, and this is the case in several groups of marine parasites. For instance, one commonly finds two or more congeneric species of Monogenea and Copepoda on one fish individual, particularly in the warm Pacific. These species frequently inhabit different microhabitats. Genetic studies may cast light on how they have originated.

Phylogeny

In recent years, some fascinating discoveries in systematic and phylogenetic zoology have been made in the sea, that is discoveries of "living fossils", forms which were thought to have been extinct for tens or hundreds of millions of years. Among these are the ancient crossopterygian fish, *Latimeria*, close to the root of all terrestrial vertebrates, and the primitive snail, *Neopilina*, belonging to a group from which all other groups of snails have evolved. Most animal groups originated in the sea, which explains why the most ancient "living fossils" are marine. Because the vast majority of marine invertebrates probably is parasitic, it is likely that studies of such forms will lead to the discovery of other ancient and primitive forms. One such form was studied on the Great Barrier Reef by Rohde (1972b,d, 1973, 1975b). He found an aspidogastrean trematode, *Lobatostoma manteri*, with the simplest obligatory two host life cycle of any trematode known. Whereas most trematodes have complicated life cycles involving at least one vertebrate and one mollusc host with several types of larval stages, in this species the snail host ingests eggs containing infective larvae, the larvae hatch in the stomach of the snail and grow up to full body size but do not produce eggs in the ducts of the digestive gland. The fish *Trachinotus blochi* becomes infected by eating snails containing the larvae. The complex life cycles of other trematodes can be derived from this simple life cycle. This conclusion was supported by the later discovery of a marine aspidogastrean, *Rugogaster hydrolagi*, by Schell (1973), which is in some respects intermediate between digenetic trematodes and aspidogastreans.

Zoogeography

The discussion of the zoogeography and latitudinal gradients of marine parasites has shown how limited our knowledge is. Detailed studies of various parasite groups in different seas may yield important conclusions concerning absolute speciation rates at different latitudes, as well as speciation rates relative to those of hosts, interchange of parasite faunas between different seas, origin of host groups, etc. Little is known about latitudinal gradients in species diversity of most groups of marine parasites, and we know even less about diversity gradients from shallow to deep water. Knowledge of such gradients may permit a test of the hypothesis that temperature is the predominant factor in determining evolutionary speed and thus species diversity.

Ecology

The discussion of parasite–parasite interactions has shown that there is little evidence for niche saturation by parasites, but much support for the view that many niches are empty and competition is of little importance. More studies are needed to evaluate the relative importance of competition, reinforcement of reproductive barriers, enhancement of intraspecific contact and possibly other factors in determining niche width and segregation. As pointed out by Rohde (1981a), ectoparasites of marine fishes are an almost ideal model to study such questions for several reasons, among which are differences in species numbers of parasite communities particularly at different latitudes, low environmental variability compared with freshwater and terrestrial environments, opportunity to examine the body surface and gills of fish accurately and quantitatively in a short time, and easy availability of fish in large numbers. In particular, too little attention has been paid to relative effects of competition and reinforcement of reproductive barriers. Parasites in the same microhabitat (for instance on the gills of fish) which are known to use the same resources (space for attachment and blood) but differ in their phylogenetic relationships may differ in their degree of niche segregation and thus may give a clue as to which of the two factors is more important (Rohde 1980d). More data are needed on niche width in parasite communities consisting of few or many species, and experiments should be conducted to show changes, if any, of niches in the presence of other species. Experiments are necessary also to evaluate the factors which determine the niches of parasites (substratum, chemical and behavioural factors, etc.). Practically nothing is known about how parasites affect host populations and especially whether and how they contribute to mass mortalities, and we still do not know what the ecological significance of cleaning symbiosis is. Also, considerable data have been accumulated on how hosts affect individual parasites, but little is known on how host reactions affect parasite populations.

Marine parasites as biological control agents

Parasites are increasingly used in attempts to control terrestrial and freshwater pests such as insects, but little work has been done on biological control in the marine environment. An exception is the work by Cooley (1962), who examined whether infections with the trematode, *Parorchis acanthus*, could be used as a means of biological control of the oyster drill, the snail *Thais haemastoma*, which may cause great damage to oysters. However, infection rates were low in the snails as well as in gulls, which serve as final hosts of the parasite, and no effective way of spreading the parasite was found.

In summary, marine parasites are a vast and little known group and present an exciting challenge to biologists interested in phylogenetic—taxonomic, ecological, zoogeographical and genetic—evolutionary problems. They are not only of theoretical interest but have great practical importance as agents of disease in marine animals, as indicators of host population patterns and migrations, and especially as ecological agents present in large numbers in all habitats in the sea.

Appendix: The Animal Kingdom

Below the level of phylum, only those protozoan groups are listed which appear in the text.

Subkingdom PROTOZOA (single celled animals)

Phylum Sarcomastigophora (flagellates and amoebas)
    Subphylum Mastigophora (flagellates)
        Class Phytomastigophorea (plant-like flagellates)
            Orders: Dinoflagellida (dinoflagellates), etc.
        Class Zoomastigophorea (animal-like flagellates)
            Orders: Kinetoplastida (trypanosomes and related forms), etc.
    Subphylum Opalinata (opalinas)
    Subphylum Sarcodina (sarcodines, amoebas)
Phylum Labyrinthomorpha
Phylum Apicomplexa (sporozoans)
    Class Sporozoea
        Subclass Gregarinia (gregarines)
        Subclass Coccidia (coccidians)
Phylum Microspora (microsporans)
Phylum Ascetospora (ascetosporans, haplosporidians)
Phylum Myxozoa (myxozoans)
Phylum Ciliophora (ciliates)
    Subclass Suctoria (suctorians) etc.

Subkingdom METAZOA (multicellular animals)

Phylum Mesozoa (mesozoans)
    Class Dicyemida (dicyemids)
    Class Orthonectida (orthonectids)

Phylum Porifera (sponges)
    Class Calcarea (calcareous sponges)
    Class Hexactinellida (glass sponges)
    Class Demosponges (horny sponges)
Phylum Cnidaria
    Class Hydrozoa (hydrozoans)
    Class Scyphozoa (jellyfish)
    Class Anthozoa (see anemones, corals, gorgonians, etc.)
Phylum Ctenophora (comb jellies)
Phylum Platyhelminthes (flatworms)
    Class Turbellaria (free-living flatworms)
    Class Trematoda (flukes)
    Class Monogenea (monogeneans, ectoparasitic flukes)
    Class Cestoda (tapeworms)
      Subclass Cestodaria (cestodarians)
      Subclass Eucestoda (genuine tapeworms)
    Class Gnathostomulida (gnathostomulids)
Phylum Priapulida (priapulids)
Phylum Entoprocta (entoprocts, nodding heads)
Phylum Nemertina (nemerteans, nemertines, ribbon worms)
Phylum Nemathelminthes (nemathelminths)
    Class Rotatoria (rotifers)
    Class Gastrotricha (gastrotrichs)
    Class Nematoda (roundworms)
    Class Nematomorpha (horsehair worms, hairworms)
    Class Kinorhyncha (kinorhynchs)
    Class Acanthocephala (spiny-headed worms, thorny-headed worms)
Phylum Annelida (segmented worms, annelids)
    Class Polychaeta (bristle worms)
    Class Myzostomida (myzostomes)
    Class Oligochaeta (earthworms)
    Class Hirudinea (leeches)
Phylum Onychophora (onychophorans)
Phylum Tardigrada (tardigrades, water bears)
Phylum Pentastomida (pentastomids, tongue worms)
Phylum Arthropoda (arthropods)
    Class Xiphosura (horseshoe crabs)
    Class Arachnida (arachnids, spider-like animals)
    Class Pycnogonida (sea spiders)
    Class Crustacea (crustaceans)
    Class Myriapoda (centipedes, millipedes etc.)
    Class Insecta (insects)

Phylum Tentaculata
    Class Phoronidea (phoronids)
    Class Bryozoa = Ectoprocta (bryozoans, moss animals)
    Class Brachiopoda (lamp shells)
Phylum Mollusca (molluscs)
    Class Solanogastres (solanogasters)
    Class Placophora (placophorans)
    Class Gastropoda (gastropods, snails)
        Subclass Prosobranchia (prosobranchs)
        Subclass Opisthobranchia (opisthobranchs)
        Subclass Pulmonata (pulmonates)
    Class Scaphopoda (tooth shells)
    Class Bivalvia (bivalves, lamellibranchs)
    Class Cephalopoda (cephalopods, octopods, squids, cuttlefish, etc.)
Phylum Echiurida (echiurans, spoon worms)
Phylum Sipunculida (sipunculans, peanut worms)
Phylum Hemichordata (hemichordates)
    Class Enteropneusta (acorn worms)
    Class Pterobranchia (pterobranchs)
Phylum Echinodermata (echinoderms)
    Class Crinoidea (sea lilies and feather stars)
    Class Holothuroidea (sea cucumbers)
    Class Echinoidea (sea urchins and sand dollars)
    Class Asteroidea (sea stars, starfish)
    Class Ophiuroidea (brittle stars)
Phylum Pogonophora (pogonophorans)
Phylum Chaetognatha (arrow worms)
Phylum Chordata (Chordates)
  Subphylum Urochordata (tunicates)
    Class Ascidiacea (ascidians, sea squirts)
    Class Thaliacea (salps)
    Class Larvacea (larvaceans, appendicularians)
  Subphylum Cephalochordata (amphioxus, lancelets)
  Subphylum Vertebrata (vertebrates)

Glossary

Some terms have several meanings. The meanings given here are only those used in this book.
Abdomen: the posterior section of the body.
Abiotic: non-living, physical and chemical.
Abscess: a collection of pus in the tissue.
Acanthella: larval stage of an acanthocephalan which develops from the acanthor.
Acanthor: first larval stage of an acanthocephalan.
Acid mucopolysaccharid: a mucopolysaccharid that contains an acid (*see* mucopolysaccharid).
Acid phosphatase: phosphatase active in acid media (*see* phosphatase).
Adhesion: abnormal union of an organ or part of an organ with some other part by formation of fibrous tissue.
Adult parasite: an organism parasitic as an adult.
Afferent branchial arteries: arteries which bring blood to the gills.
Agamete: daughter cell which has originated without fertilization.
Aggregated distribution: spatial distribution in which organisms are clustered within their area of distribution (*see* clustered distribution, overdispersed distribution, contagious distribution).
Alkaline phosphatase: phosphatase active in alkaline media (*see* phosphatase).
Allopatric speciation: species formation by geographical isolation (*see* geographic speciation).
Ambulacral groove: furrow on the oral surface of echinoderms with rows of tube feet.
Amino acid: any acid containing the amino group, chief component of protein.
Amoebocyte: a cell capable of amoeba-like locomotion.
Amorphic: structureless.

Glossary

Amphiboreal: having a northern temperate distribution, both in the Atlantic and the Pacific Oceans.
Anisakiasis: disease caused by *Anisakis*.
Antibody: specific protein (globulin) formed by the host as a reaction to an antigen.
Antiboreal: distributed in the southern temperate ocean.
Antigen: a substance which induces the formation of antibodies.
Anting: 1) passive anting: bathing of birds in ants;
2) active anting: activity by which birds put ants under their feathers with their beak.
Asexual reproduction: reproduction without male and female cells.
Atrium: 1) the anterior cavity of sponges;
2) the cavity between the pharynx and the body wall in tunicates;
3) blood chamber that receives blood from the veins and pumps blood into the ventricle of vertebrates.
Autolysis: process of self-digestion in animal tissue.
Bathypelagic: pertaining to the dark middle zone of the ocean, with temperatures between $10°$ and $4°C$, approximately 800–4,000 m deep.
Benthic: living on the sea floor.
Benthopelagic: pertaining to organisms close to the sea floor.
Binary fission: cell division into two daughter cells.
Biotic: living.
Blastula: an early stage in the development of the embryo consisting of a sphere of cells, either hollow or solid.
Boreal: distributed in the northern temperate ocean.
Bothridium: attachment organ of some tapeworms.
Branchial chamber: gill chamber.
Bursa: a posterior organ of male nematodes and acanthocephalans holding the female during copulation.
Capsule: a layer of connective tissue fibres formed by a host around a parasite.
Carrying capacity: number of individuals which an environment can support.
Caudal: pertaining to the tail.
Cement gland: a gland which produces a secretion serving for attachment.
Cephalic papilla: head papilla.
Cercaria: tailed larva of a trematode.
Character displacement: increased differences between two species in the area where they occur together.

Cholangio hepatitis: inflammation of the liver and bile ducts.
Chronic: long lasting (disease).
Cleaning symbiosis: Association between a cleaner organism (which feeds on parasites, diseased tissue and dirt) and a host (whose parasites etc. are removed).
Cleavage: cell divisions leading to the formation of the embryo.
Clustered distribution: spatial distribution in which organisms are clustered within their area of distribution (*see* aggregated distribution, overdispersed distribution, contagious distribution).
Coelom: body cavity lined by mesoderm (peritoneum).
Colony: permanent association of loosely connected cells or individuals.
Commensalism: an association in which one partner uses food supplied in the internal and external environment of the other, without affecting it in any way.
Congeneric: belonging to the same genus.
Connective tissue: tissue with much extracellular substance, connecting and supporting epithelial, muscular and nervous tissues.
Contagious distribution: spatial distribution in which organisms are clustered within their area of distribution (*see* aggregated distribution, overdispersed distribution, clustered distribution).
Coracidium: the ciliated larva of many tapeworms.
Cyst: a protective sheath secreted by a parasite.
Cystacanth: juvenile stage of some acanthocephalans which becomes the adult when eaten by the final host.
Cytochemistry: chemistry of the cell.
Definitive host: the host in which a parasite reaches sexual maturity (*see* final host).
Density dependent: dependent on the population density (individuals per unit area).
Density independent: independent of the population density (individuals per unit area).
Density of infection: the total number of parasite individuals divided by the total number of host individuals examined, mean infection intensity including zeros (*see* intensity of infection).
Dermatitis: inflammation of the skin.
Dermis: the layer of connective tissue beneath the surface epithelium in nemerteans.
Detritus: dead particles.

Devonian: pertaining to the geologic era of the Paleozic between the Silurian and Carboniferous, lasting from circa 400 to circa 350 million years ago.
Direct life cycle: life cycle not involving an alternation of hosts.
Dorsoventral: extending from the dorsal to the ventral side.
Ecological host specificity: restriction of the host range to some hosts living in the same habitat or in similar habitats.
Ecology: science dealing with the relations between organisms and their environment.
Ectoparasite: a parasite on the surface of a host.
Endemic: confined to, or indigenous in, a certain region.
Endogenous: originating from within a cell or organism.
Endoparasite: a parasite in the interior of a host.
Enzyme: a complex organic substance that accelerates specific chemical transformations.
Eosinophilic meningoencephalitis: inflammation of the brain and its membranes characterized by abnormally large number of eosinophils (type of white blood cells).
Epidemic: a rapidly spreading disease affecting large numbers of organisms (usually applies to humans, *see* epizootic).
Epidemiology: the science of epidemics
Epidermis: layer(s) of cells on the surface of an organism.
Epithelium: layer(s) of cells lining the surface of an organism or its cavities.
Epizoon: an organism living on the surface of an animal.
Epizootic: a disease attacking a large number of animals at the same time (*see* epidemic).
Erythrocyte: a red blood cell.
Essential antigen: an antigen inducing the formation of protective antibodies (*see* protective antibody, antigen).
Esterase: enzyme that accelerates the decomposition of certain chemical compounds, the esters.
Even distribution: spatial distribution in which the distances between organisms are more or less equal.
Extrinsic: interspecific, due to actions between species.
Exudate: a discharge of fluid through a pore or wound.
Facultative parasite: a parasite which can also live without a host.
Fatty acid: saturated acids, some of which occur in fats.
Fibroblast: a connective tissue cell.
Fibrosis: an increase of fibrous (connective) tissue.

Final host: the host in which a parasite reaches sexual maturity (*see* definitive host).
Frequency of infection: percent of host individuals infected (*see* prevalence).
Functional antigen: an antigen inducing the formation of protective antibodies (*see* essential antigen).
Fundamental niche: niche in the absence of competitors.
Gametocyte: a cell which develops into a gamete or several gametes.
Gastritis: inflammation of the stomach.
Gastrovascular cavity: a cavity with the combined function of a digestive tract and a vascular (blood) system.
Generation time: time from formation of the egg to development of gravid female.
Geographical speciation: species formation by geographical isolation (*see* allopatric speciation).
Gigantism: state of being abnormally large.
Glucose: a sugar, type of carbohydrate.
Glycogen: a type of carbohydrate related to starch.
Glycoprotein: a protein that contains a carbohydrate group other than nucleic acid.
Granulocyte: white blood cells with granules.
Granuloma: a nodular lesion of inflammatory tissue with significant granular tissue formation.
Guild sign: markings characteristic of several species of animals not indicating their taxonomic status but their ecological function (for example cleaning).
Gustatory: pertaining to taste.
Haemocoel: spaces in the tissue filled with blood.
Haemoglobin: red respiratory pigment containing iron.
Haemolymph: body fluid of molluscs and arthropods.
Haemorrhage: discharge of blood from the blood vessels, caused by injury.
Helminth: parasitic worm.
Hermaphrodite: an individual with male and female reproductive organs.
Hexamitiasis: disease caused by *Hexamita*.
Histochemistry: the science of tissue chemistry.
Host range: the number of host species infected by a parasite species.
Host specificity: restriction to certain hosts.
Humoral: pertaining to body fluid.
Hyalinocyte: white blood cells without granules.
Hyperparasite: a parasite of a parasite.

Hyperplasia: an increase in the number of cells.
Hypertrophy: overdevelopment, increase in the size of an organ or cell.
Hypodermis: a cellular layer below the cuticle (in nematodes) or below the exoskeleton (in arthropods).
Immune: non-receptive to infection with a parasite as a result of previous exposure to it.
Immunity: state of being immune (*see* immune).
Immunoglobulin: protein with known or potential antibody activity.
Index parasite: parasite indicating some ancient condition.
Indirect life cycle: life cycle with an alternation of hosts.
Inflammation: a local tissue response to cellular injury marked by widening of blood capillaries, infiltration by white blood cells, etc.
Intensity of infection: number of parasites per host individual (*see* density of infection).
Intention movement: a characteristic (species specific) movement of an animal that is initiated but not completed.
Intermediate host: a host harbouring the immature and developing stage of a parasite.
Intertidal zone: the coastal zone between low and high water.
Intraspecific parasite: an organism parasitizing an organism of the same species.
Intrinsic: intraspecific, due to actions between individuals of one species.
Intrinsic rate of population growth: rate of population increase entirely due to the reproductive potential of the population.
Invertebrate: an animal without a backbone.
Juvenile: young, immature stage.
K-selection: selection which favours maintenance of the individual but not production of many offspring (*see* r-selection, K-strategy).
K-strategy: ecological strategy with a tendency towards small numbers of offspring and great complexity of individuals (*see* K-selection).
Labial papilla: a papilla on the lips.
Larval parasite: an organism parasitic as a larva.
Latent parasite: a parasite not causing symptoms.
Leucocyte: a colourless blood corpuscle.
Leucocytosis: increase in the number of white blood cells.
Lipid: a fat or related substance.
Littoral: pertaining to the sea shore.

Lymph: a more or less colourless body fluid (in vertebrates contained in the lymph vessels which are separate from the blood vessels but in close contact with them).
Macrogamete: female sex cell of many Protozoa.
Macrohabitat: the habitat of a parasite's host.
Macrophage: large cell capable of phagocytosis (*see* phagocytosis).
Mandible: feeding appendage of Crustacea.
Mechanoreceptor: a receptor sensitive to mechanical stimuli.
Medusa: the free-swimming stage of a cnidarian (jellyfish).
Merozoite: a daughter cell of a schizont in the Apicomplexa (*see* schizont).
Mesopelagic: living in midwater 100–800 m deep.
Metacercaria: encysted larval stage (cercaria) of a trematode.
Metaplasia: a transformation of one tissue type into another.
Metazoa: multicellular animals.
Microgamete: the male sex cell of many Protozoa.
Microhabitat: the habitat of a parasite within or on a host.
Midwater: the zone of the oceans from below the surface zone (approx. 100 m deep) to above the bottom layer.
Mimic: an organism imitating another, an organism resembling another.
Mucopolysaccharid: a mucus-like substance that is not connected to a protein.
Mucosa: mucous membrane, surface layer in the intestine of vertebrates.
Mucus: a viscid, slippery secretion.
Multiple fission: cell division into many daughter cells.
Mutation: a change in the genes or chromosomes of an organism.
Mutualism: an association between two organisms beneficial to both partners.
Nauplius larva: 1st larval stage of crustaceans, with three pairs of appendages.
Necrosis: death of tissue.
Nephridia: a tubular excretory organ which opens to the exterior through an excretory pore.
Neutral lipid: a lipid which is neither acidic nor basic (*see* lipid).
Niche: the total of the relations of an organism to its environment, an organism's place in nature.
Obligatory parasite: a parasite which cannot live without a host.
Oedema: abnormal accumulation of fluid in tissue spaces or body cavities.
Oncomiracidium: the ciliated larva of a monogenean.
Oocyst: the fertilized egg cell of Apicomplexa surrounded by a cyst wall in which the sporozoites develop.

Opisthaptor: the posterior attachment organ of monogeneans.
Ovary: the organ producing egg cells.
Overdispersed distribution: spatial distribution in which organisms are clustered within their area of distribution (*see* aggregated distribution, clustered distribution, contagious distribution).
Oviger: the egg-carrying limb of a pycnogonid.
Pancreas: gland of most vertebrates discharging into the intestine.
Panthothenate: component of vitamin B_2.
Parapodia: limb-like appendages of a polychaete or myzostomid.
Parasitic castration: sterility caused by parasites.
Parasitism: an association between two organisms in which one, the parasite, depends on the other, the host, for some essential factor(s).
Parasitoid: an animal which as a larva feeds for an extended period on another living animal and finally kills it.
Paratenic host: a host in which there is no development of the immature parasite (*see* transport host).
Parenchyma: cells of an organ that perform the specific function of the organ.
Parthenogenesis: reproduction by the development of an unfertilized egg cell.
Pathogen: pathogenic organism, organism causing disease.
Pathology: science of disease.
Pelagic: pertaining to organisms that live in the free water.
Pellicle: the elastic thickened surface membrane of some flagellates and of the ciliates.
Pericardium: the cavity surrounding the heart in arthropods.
Periodic parasite: a parasite visiting its host(s) at intervals.
Permanent Parasite: a parasite permanently in contact with its host.
Phagocyte: a cell capable of phagocytosis.
Phagocytosis: uptake of solid material by a cell.
Pharynx: a muscular organ for ingesting food.
Phasmid: unicellular chemoreceptor in the tail region of nematodes.
Phoresis (=phoresy): an association in which one organism uses another as a means of transport (or protection).
Phosphatase: enzyme that accelerates decomposition of certain phosphorus containing compounds.
Phospholipid: a lipid that contains phosphorus (*see* lipid).
Photosynthesis: formation by plants of carbohydrates with the aid of light.

Phototaxis: a directed reaction to light.
Phylogenetic host specificity: restriction of the host range to some related host taxa.
Phylogeny: the history of a taxon.
Planula: the ciliated larva of a cnidarian.
Plankton: organisms floating passively in the water or with swimming movements not strong enough to overcome the effect of currents.
Plasma: the fluid part of blood and lymph.
Plasmodium: a mass of cytoplasm containing several nuclei.
Plastid: a cell component (organelle) containing a pigment capable of photosynthesis.
Pleistocene: glacial, or geologic era lasting from circa 1 million to circa 10,000 years ago.
Plerocercoid: a larval stage of many tapeworms.
Pleurobranch: a small gill of teleost fish (sometimes rudimentary or missing) attached to the wall of the mouth cavity.
Polar capsule: a capsule in the spore of certain Protozoa containing the polar filament (*see* polar filament).
Polar filament: a thread-like structure in the polar capsule of the spore of certain Protozoa (*see* polar capsule).
Polyp: the sessile stage in the life cycle of cnidarians, typically consisting of a tube-like body and upward-directed mouth opening and tentacles.
Praniza: larval stage of the isopod group Gnathiidea.
Predation: living by preying upon other animals, generally killing them.
Prevalence: percent of host individuals infected (*see* frequency of infection).
Proboscis: an extendable anterior organ for attachment (for example Acanthocephala) or catching prey (Nemertina).
Procercoid: a larval stage of many tapeworms.
Progenesis: maturation at the larval stage.
Protective antibody: an antibody responsible for protecting a host against further infection (*see* antibody).
Proteolipid: a compound that consists of protein and lipid (*see* lipid).
Proteolytic: capable of splitting proteins.
Protonymphon: larval stage of most Pycnogonida (sea spiders).
Proximate cause: the conditions of the environment or internal physiology that trigger the response of an organism.
Pseudocleaning: the stimulation of fish by sessile organisms which

190 Glossary

resembles stimulation by cleaners but does not lead to cleaning.
Purine: a chemical compound, necessary for the synthesis of nucleic acids.
Pyloric caeca: tubular pouches that open from the posterior (pyloric) end of the stomach of fishes into the gut.
Pyridoxine: component of vitamin B_6.
Random distribution: spatial distribution in which the distances between organisms are random.
Realized niche: niche compressed by competitors.
Receptor: a sense organ or sensory cell.
Redia: a larval stage of trematodes which produces cercariae (*see* cercaria).
Relict fauna: a fauna which has been isolated from its parent fauna for a long time (usually as a result of contraction of a formerly widely distributed fauna).
Respiratory tree: a respiratory system of tubules in sea cucumbers (Holothuria) connected to the cloaca.
Rheoreceptor: a receptor sensitive to currents.
Rheotaxis: directed reaction to fluid currents.
Ring canal: part of the water-vascular (= ambulacral) system of echinoderms forming a central ring.
Rosette: the posterior attachment organ of some tapeworms (Gyrocotylidae).
r-selection: selection which favours production of many offspring at the expense of maintaining the individual (*see* K-selection, r-strategy).
r-strategy: an ecological strategy with a tendency towards rapid production of large numbers of offspring and low complexity of individuals (*see* r-selection).
Scanning electron microscope: a microscope using a beam of electrons sweeping over the surface of an object and producing a plastic image of the surface.
Schizogony: asexual part of the life cycle of Apicomplexa which leads to the production of several merozoites and is followed by sexual reproduction.
Schizont: a cell in the life cycle of Apicomplexa which divides into several merozoites.
Scolex: the anterior attachment organ of some tapeworms.
Seta: bristle.
Scyphomedusa: the medusa of Scyphozoa, jellyfish.
Selfer: an organism capable of self-fertilization.

Sequential hermaphroditism: hermaphroditism in which one sex matures before the other (*see* hermaphrodite).
Silurian: pertaining to the geologic era of the Paleozoic between the Ordovician and Devonian, lasting from circa 430 to circa 400 million years ago.
Simultaneous hermaphroditism: hermaphroditism in which both sexes are mature at the same time (*see* hermaphrodite).
Sinus: large intercellular space.
Spore: 1. oocyst of many Apicomplexa (*see* oocyst);
2. body formed in the oocyst and containing the sporozoites of Apicomplexa.
Sporocyst: 1. the larval stage of flukes which develops from the ciliated first larva, the miracidium;
2. the cyst surrounding the sporozoites in some Apicomplexa.
Sporogony: part of the life cycle of Apicomplexa preceded by fertilization of an egg cell, in which sporozoites are formed within a cyst.
Sporozoite: cells formed in the oocyst of Apicomplexa.
Stolon: modified parts of the colonies of many bryozoans and ascidians from which new colonial animals arise by budding and which connect the animals forming the colony.
Strobila: the chain of segments of a tapeworm.
Sublittoral: the zone below the littoral (*see* littoral).
Submucosa: the layer in the intestine of vertebrates beneath the mucosa, consisting mainly of connective tissue.
Symbiosis: 1. an association which is beneficial to and obligatory for the two partners;
2. any close association between two organisms.
Sympatric speciation: species formation within one geographical area (without geographical isolation).
Tangoreceptor: a receptor sensitive to touch.
Taxon: a taxonomic (systematic) unit, for example species, genus, family, order.
Temporary parasite: a parasite which is in contact with its host for short period(s) only.
Thiamine: vitamin B_1.
Thorax: the middle region of the body of arthropods.
Transport host: a host that carries a non-developing parasite to another host (*see* paratenic host).
Trypanosomiasis: disease caused by trypanosomes.

Ulcer: a superficial sore discharging pus.
Ulceration: process of forming an ulcer, state of being ulcerated (*see* ulcer).
Ultimate cause: the environmental conditions that lead to the evolution of a response of an organism.
Urea: a water soluble constituent of urine containing nitrogen.
Ureter: duct between kidney and bladder or cloaca in vertebrates.
Vascular system: a tubular system filled with liquid.
Vasodilation: widening of blood vessels.
Vector: a host which carries an infective stage of a parasite
Ventral disc: the ventral attachment organ of aspidogastrids.
Ventricle: heart chamber which delivers the blood to the arteries.
Verminous pneumonia: inflammation of the lungs due to a worm infection.
Virulence: the relative infectiousness or malignancy of a disease agent.
Vitamin: certain food constituents of which very small quantities are necessary for metabolism.
Zoochlorella: green, unicellular, symbiotic alga.
Zooplankton: the animal component of plankton (*see* plankton).
Zooxanthella: brown, unicellular, symbiotic alga.

References

Abel, E.F. 1971. "Zur Ethologie von Putzsymbiosen einheimischer Süsswasserfische im natürlichen Biotop." *Oecologia* 6: 133–51.
Achmerov, A.Kh. 1954. "(A new series of *Dactylogyrus* from the gills of Amurean *Hypophthalmichthys molitrix* (Val.).)" *Dokl. Akad. Nauk SSSR* 98: 167–68 (in Russian).
Anderson, D.T., Fletcher, M.J. and Lawson-Kerr, C. 1976. "A marine caddis fly, *Philanisus plebeius*, ovipositing in a starfish, *Patiriella exigua*." *Search* 7: 483–84.
Anderson, D.T. and Lawson-Kerr, C. 1977. "The embryonic development of the marine caddis-fly, *Philanisus plebeius* Walker (Trichoptera: Chathamidae)." *Biol. Bull.* 153: 98–105.
Anderson, R.M. 1974. "An analysis of the influence of host morphometric features on the population dynamics of *Diplozoon paradoxum* (Nordmann, 1832)." *J. Anim. Ecol.* 43: 873–87.
Anderson, R.M. 1976. "Dynamic Aspects of Parasite Population Ecology." In: Kennedy, C.R. ed: *Ecological Aspects of Parasitology*. Amsterdam: North-Holland Publ. Co. 431–61.
Anderson, R.M. 1980. "Depression of host population abundance by direct life cycle macroparasites." *J. Theor. Biol.* 82: 293–311.
Anderson, R.M. and May, R.M. 1979. "Population biology of infectious diseases: Part I." *Nature* 280: 361–67.
Andrews, J.D. 1968. "Oyster mortality studies in Virginia. VII. Review of epizootiology and origin of *Minchinia nelsoni*." *Proc. Nat. Shellfish Assoc.* 58: 23–36.
Andrews, J.D. and Wood, J.L. 1967. "Oyster mortality studies in Virginia. VI. History and distribution of *Minchinia nelsoni*, a pathogen of oysters, in Virginia." *Chesapeake Science* 8: 1–13.
Andrews, J.D., Wood, J.L. and Hoese, H.D. 1962. "Oyster mortality studies in Virginia: III. Epizootiology of a disease caused by *Haplosporidium costale* Wood and Andrews." *J. Insect. Pathol.* 4: 327–43.
Ankel, F. and Christensen, A.M. 1963. "Non-specificity in host selection by *Odostoma scalaris* MacGillivray." *Videnskabelige Meddelelser Dansk Naturh. For. København* 125: 321–25.
Arme, C. 1976. "Feeding." In: Kennedy, C.R. ed.: *Ecological Aspects of Parasitology*. Amsterdam, Oxford: North Holland Publ. Co., 75–97.

Arme, C. and Fox, M.G. 1974. "Oxygen uptake by *Diclidophora merlangi* (Monogenea)." *Parasitol.* 69: 201–205.

Armitage, J.G. 1980. "The taxonomy, zoogeography and site specificity of the metazoan ectoparasites of the snapper, *Chrysophrys auratus*." M.Sc. Prelim. Thesis. University of New England, Armidale.

Armstrong, H.W. 1975. "A study of the helminth parasites of the family Macrouridae from the Gulf of Mexico and Caribbean Sea: their systematics, ecology and zoogeographical implications." *Dissertation Abs. Inter.* 36B: 123–24.

Arndt, W. 1940. "Der prozentuelle Anteil der Parasiten auf und in Tieren im Rahmen des aus Deutschland bisher bekannten Tierartenbestandes." *Z. Parasitenk.* 11: 684–89.

Arundel, J.H. 1978. "Parasites and parasitic diseases of Australian marine mammals." *Proc. No. 36. Refresher course. Veterinarians. Fauna—Part B*, 6–10, 1978, Sydney, 323–33.

Ashizawa, H., Ezaki, T., Arie, Y., Nosaka, D., Tateyama, S. and Owada, K. 1978. "(Pathological findings on lungworm disease in California sea lions.)" *Bull. Fac. Agriculture Miyazaki Univ.* 25: 287–95 (in Japanese).

Atkins, P.D. and Gorlick, D.L. 1976. "Effects of the cleaning behaviour of *Labroides dimidiatus* on fish ectoparasites." *Pacific Sci.* 30: 214.

Bacha, W.J. jr. 1966. "Viable egg production in *Zygocotyle lunata* following monometacercarial infections." *J. Paras.* 52: 1216–17.

Baer, J.G. 1952. *Ecology of Animal Parasites*. Urbana: Univ. Illinois Press.

Baer, J.G., Miranda, H., Fernandez, W. and Medina, J. 1967. "Human diphyllobothriasis in Peru." *Z. Parasitenk.* 28: 277–89.

Baker, R.O. 1977. "An ecological study of parasitism in estuarine fishes." *Dissertation Abs. Inter.* 378: 4272.

Bang, F.B. 1970. "Disease mechanisms in crustacean and marine arthropods." In: Sniezko, S.F. ed.: *A Symposium on Diseases of Fishes and Shellfishes*, Amer. Fish. Soc., Special Publ. 5, Washington, D.C., 383–404.

Banning, P. Van, and Becker, H.B. 1978. "Long-term survey data (1965–1972) on the occurrence of *Anisakis* larvae (Nematoda: Ascaridida) in herring, *Clupea harengus* L., from the North Sea." *J. Fish Biol.* 12: 25–33.

Barnes, R.D. 1974. *Invertebrate Zoology*. 3rd ed. Philadelphia, London, Toronto: W.B. Saunders.

Barrington, E.J.W. 1957. "The alimentary canal and digestion." In: Brown, M.E. ed.: *The Physiology of Fishes*, vol. 1. New York: Academic Press, 109–61.

Barry, J.M. and O'Rourke, F.J. 1959. "Species specificity of fish mucus." *Nature* 184: 2039.

Baudoin, M. 1975. "Host castration as a parasitic strategy." *Evolution* 29: 335–52.

Baylis, H.A. 1947. "Some roundworms and flatworms from the West Indies and Surinam II. Cestodes." *J. Linn. Soc.* 41: 407–13.

Bayne, B.L., Gee, J.M., Davey, J.T. and Scullard, C. 1978. "Physiological

responses of *Mytilus edulis* L. to parasitic infestation by *Mytilicola intestinalis*." *J. Cons. int. Explor. Mer.* 38: 12–17.
Bearup, A.J. 1955. "A schistosome larva from the marine snail *Pyrazus australis* as a cause of cercarial dermatitis of man." *Med. J. Austr.* 1: 955–60.
Bearup, A.J. 1956. "Life cycle of *Austrobilharzia terrigalensis* Johnston 1917." *Parasitol.* 46: 470–79.
Becker, C.D. 1970. "Haematozoa of fishes, with emphasis on North American records." In: Sniezko, S.F. ed.: *A Symposium of Diseases of Fishes and Shellfishes.* Amer. Fish. Soc., Special Publ. 5, Washington, D.C., 82–100.
Becker, J. 1942. "Die Abwehreinrichtungen von Haut und Kieme beim Karpfen gegenüber mechanischen, chemischen und parasitären Reizen." *Int. Revue ges. Hydrobiol.* 41: 265–344.
Befus, A.D. and Oodesta, R.B. 1976. Intestine. In: Kennedy, C.R. ed.: *Ecological Aspects of Parasitology.* Amsterdam, Oxford: North Holland Publ. Co., 303–25.
Berland, B. 1980. "Are parasites always harmful?" *Proc. EMOP.* III. Cambridge, 202.
Bishop, L. 1979. "Parasite-related lesions in a bearded seal, *Erignathus barbatus*." *J. Wildlife Diseases* 15: 285–93.
Bishop, R.K. and Cannon, L.R.G. 1979. "Morbid behaviour of the commercial sand crab, *Portunus pelagicus* (L.), parasitized by *Sacculina granifera* Boschma, 1973 (Cirripedia: Rhizocephala)." *J. Fish. Dis.* 2: 131–44.
Bloom, B.R. 1979. "Games parasites play: how parasites evade immune surveillance." *Nature* 279: 21–26.
Bogdanova, E.A. 1957. "(On certain aspects of the biology of *Dactylogyrus skrjabini* Achmerov, 1954 – Parasite of Amur *Hypophthalmichthys*.)" *Dokl. Akad. Nauk SSSR* 113: 1391–93 (in Russian).
Bolster, G.C. 1954. "The biology and dispersal of *Mytilicola intestinalis* Steuer, a copepod parasite of mussels." *Fisheries Investig. Min. Agric. Fish.*, ser. II, 18: 1–30.
Bortone, S.A., Bradley, W.K. and Oglesby, J.L. 1978. "The host–parasite relationship of two copepod species and two fish species." *J. Fish Biol.* 13: 337–50.
Boss, K.J. and Merrill, A.S. 1965. "Degree of host specificity in two species of *Odostomia* (Pyramidellidae: Gastropoda)." *Proc. malac. Soc. Lond.* 36: 349–55.
Boxshall, G.A. 1974. "Infections with parasitic copepods in North Sea marine fishes." *J. mar. biol. Ass. U.K.* 54: 355–72.
Boxshall, G.A. 1976. "The host specificity of *Lepeophtheirus pectoralis* (Müller, 1776) (Copepoda: Caligidae)." *J. Fish Biol.* 8: 255–64.
Brandal, P.O., Egidius, E. and Romslo, I. 1976. "Host blood: a major component for the parasitic copepod *Lepeophtheirus salmonis* Kröyeri, 1838 (Crustacea: Caligidae)." *Norw. J. Zool.* 24: 341–43.
Bremer, H. 1972. "Einige Untersuchungen zur Histochemie der sezernieren-

den Elemente der Teleostier—Epidermis." *Acta histochem.* 43: 28–40.
Bresslau, E. 1928–33. "Turbellaria." In: Kükenthal, W. and Krumbach, Th. ed.: *Handbuch der Zoologie*, vol. 2. 1st half, 52–293.
Briggs, J.C. 1966. "Oceanic islands, endemism, and marine paleotemperatures." *Syst. Zool.* 15: 153–63.
Brinkmann, A. 1952. "Fish trematodes from Norwegian waters. I. The history of fish trematode investigations in Norway and the Norwegian species of the order Monogenea." *Årbok. Bergen (Naturv. rekke).* 1: 1–134.
Brown, W.L. jr. and Wilson, E.O. 1956. "Character displacement." *Syst. Zool.* 5: 49–64.
Buckner, R.L.; Overstreet, R.M. and Heard, R.W. 1978. "Intermediate hosts for *Tegorhynchus furcatus* and *Dollfusentis chandleri* (Acanthocephala)." *Proc. Helm. Soc. Washington.* 45: 195–201.
Budiarso, I.T., Palmieri, J.R., Imes, G.D. jr., Allen, J.F. and Lepes, M.M. 1979. "Two species of trematodes causing nasal lesions in dugongs." *J. Paras.* 65: 568.
Bullock, A.M., Marks, R. and Roberts, R.J. 1978. "The cell kinetics of teleost fish epidermis: mitotic activity of the normal epidermis at varying temperatures in plaice (*Pleuronectes platessa*)." *J. Zool.* 184: 423–28.
Bullock, A.M., Roberts, R.J. and Gordon, J.D.M. 1976. "A study on the structure of the Whiting integument (*Merlangius merlangus* L.)." *J. mar. biol. Ass. U.K.* 56: 213–26.
Burn, P.R. 1980. "Density dependent regulation of a fish trematode population." *J. Paras.* 66: 173–74.
Bussmann, B. and Ehrich, S. 1979. "Investigations on infestation of blue whiting (*Micromesistius poutassou*) with larval *Anisakis* sp. (Nematoda: Ascaridida)." *Archiv für Fischereiwissenschaft* 29: 155–65.
Bychowsky, B.E. 1933. "Die Bedeutung der monogenetischen Trematoden für die Erforschung der systematischen Beziehungen der Karpfenfische." *Zool. Anz.* 102: 243–51.
Bychowsky, B.E. 1957. *Monogenetic Trematodes.* English Translation. Washington: Am. Inst. Biol. Sc.
—Bychowsky, I. and Bychowsky, B. 1934. "Über die Morphologie und die Systematik des *Aspidogaster limacoides* Diesing." *Z. Parasitenk.* 7: 125–37.
Bychowsky, B.E. and Nagibina, L.T. 1967. "(On 'intermediate' hosts in monogeneans (Monogenoidea).)" *Parasitologiya* 1: 117–22 (in Russian).
Byrnes, T. 1980. "The taxonomy, site specificity and zoogeography of metazoan ectoparasites infecting the southern, yellowfin and tropical black bream." B. Sc. (Hon.). Thesis. University of New England, Armidale.
Campbell, R.A., Haedrich, R.L. and Munroe, T.A. 1980. "Parasitism and ecological relationships among deep-sea benthic fishes." *Mar. Biol.* 57: 301–313.
Canning, E.V. 1975. "Microsporidian parasites of Platyhelminthes." *CIH Misc. Publ.* No. 2. 32 pp.

Cannon, L.R.G. 1977a. "Some ecological relationships of larval ascaridoids from south-eastern Queensland marine fishes." *Int. J. Paras.* 7: 227–32.
Cannon, L.R.G. 1977b. "Some larval ascaridoids from south-eastern Queensland marine fishes." *Int. J. Paras.* 7: 233–43.
Cannon, L.R.G. 1978a. "A larval ascaridoid nematode from Queensland scallops." *Inter. J. Paras.* 8: 75–80.
Cannon, L.R.G. 1978b. "Marine cercariae from the gastropod *Cerithium moniliferum* Kiener at Heron Island, Great Barrier Reef." *Proc. R. Soc. Qd.* 89: 45–77.
Cannon, L.R.G. 1979. "Ecological observations on *Cerithium moniliferum* Kiener (Gastropoda: Cerithiidae) and its trematode parasites at Heron Island, Great Barrier Reef." *Austr. J. Mar. Freshwater Res.* 30: 365–74.
Casimir, M.J. 1969. "Zum Verhalten des Putzerfisches *Symphodus melanocercus* (Risso)." *Z. Tierpsych.* 26: 225–29.
Caullery, M. 1952. *Parasitism and Symbiosis.* English translation London: Sidgwick and Jackson.
Cheng, L. ed. 1976. *Marine Insects.* Amsterdam, Oxford: North-Holland Publ. Co.
Cheng, T.C. 1963a. "Histological and histochemical studies on the effects of parasitism of *Musculium partumeium* (Say) by the larvae of *Gorgodera amplicava* Looss." *Proc. Helminth. Soc. Washington.* 30: 101–107.
Cheng, T.C. 1963b. "Some biochemical immunological aspects of host–parasite relationships." *Annals. N.Y. Acad. Sci.* 113: 1–510.
Cheng, T.C. 1964. *The Biology of Animal Parasites.* Philadelphia–London: W.B. Saunders.
Cheng, T.C. 1966a. "Perivascular leucocytosis and other types of cellular reactions in the oyster *Crassostrea virginica* experimentally infected with the nematode *Angiostrongylus cantonensis*." *J. Invert. Path.* 8: 52–58.
Cheng, T.C. 1966b. "The coracidium of the cestode *Tylocephalum* and the migration and fate of this parasite in the American oyster, *Crassostrea virginica.*" *Trans. Amer. Microsc. Soc.* 85: 246–55.
Cheng, T.C. 1967. "Marine molluscs as hosts for symbioses with a review of known parasites of commercially important species." *Adv. Mar. Biol.* 5: 1–424.
Cheng, T.C. 1975a. "Functional morphology and biochemistry of molluscan phagocytes." In: Bulla, L.A. jr. and Cheng T.C. ed.: *Pathobiology of Invertebrate Vectors of Disease. Annals N.Y. Acad. Sc.:* 343–79.
Cheng, T.C. 1975b. "*Echinocephalus crassostreai* sp. nov., a larval nematode from the oyster *Crassostrea gigas* in the Orient." *J. Invert. Path.* 26: 81–90.
Cheng, T.C. and Burton, R.W. 1966. "Relationships between *Bucephalus* sp. and *Crassostrea virginica*: a histochemical study of some carbohydrates and carbohydrate complexes occurring in the host and parasite." *Parasitol.* 56: 111–22.
Cheng, T.C. and Rifkin, E. 1968. "The occurrence and resorption of *Tylocephalum* metacestodes in the clam *Tapes semidecussata.*" *J. Invert. Path.* 10: 65–69.

Cheng, T.C. and Rifkin, E. 1970. "Cellular reactions in marine molluscs in response to helminth parasitism." In: Snieszko, S.F. ed.: *A Symposium on Diseases of Fishes and Shellfishes. Amer. Fish. Soc., special Publ.* 5, Washington, D.C. 443–96.

Cheng, T.C., Shuster, C.N. jr. and Anderson, A.H. 1966. "A comparative study of the susceptibility and response of eight species of marine pelecypods to the trematode *Himasthla quissetensis*." *Trans. Amer. Microsc. Soc.* 85: 284–95.

Christensen, A.M. 1958. "On the life history and biology of *Pinnotheres pisum*." *Proc. 15th Int. Congr. Zool.*: 167–270.

Christensen, A.M. and Kanneworff, B. 1976. "On some cocoons belonging to undescribed species of endoparasitic turbellarians." *Ophelia* 4: 28–42.

Christensen, A.M. and McDermott, J.J. 1957. "Life-history and biology of the Oyster crab, *Pinnotheres ostreum* Say." *Biol. Bull.* 113: 146–79.

Chua, T.H. 1979. "A comparative study of the searching efficiencies of a parasite and a hyperparasite." *Res. Popul. Ecol.* 20: 179–87.

Clark, M.R. 1966. "A review of the systematics and ecology of oceanic squids." *Adv. mar. Biol.* 4: 91–300.

Clark, R.B. 1956. "*Capitella capitata* as a commensal, with a bibliography of parasitism and commensalism in the polychaetes." *Ann. Mag. Nat. Hist.* 9: 433–48.

Clarke, T.A., Flechsig, A.O. and Grigg, R.W. 1967. "Ecological studies during project Sealab II." *Science* 157: 1381–89.

Cole, H.A. and Savage, R.E. 1951. "The effect of the parasitic copepod, *Mytilicola intestinalis* (Steuer) upon the condition of mussels." *Parasitol.* 41: 156–61.

Collard, S.B. 1970. "Some aspects of host–parasite relationships in mesopelagic fishes." In: Snieszko, S.F. ed.: *A Symposium on Diseases of Fishes and Shellfishes. Amer. Fish. Soc.*, Special Publ. 5, Washington, D.C.

Comps, M. 1978. "Evolution des recherches et études récentes en pathologie des huitres." *Oceanologica Acta.* 1: 255–62.

Cooley, N.R. 1962. "Studies on *Parorchis acanthus* (Trematoda: Digenea) as a biological control for the southern oyster drill, *Thais haemastoma*." *Fish. Bull., U.S. Nat. Mar. Fish. Serv.* 62: 77–91.

Corbel, M.J. 1975. "The immune response in fish: a review." *J. Fish. Biol.* 7: 539–63.

Cordes, D.O. and O'Hara, P.J. 1979. "Diseases of captive marine mammals." *N.Z. Veterinary Journal* 27: 147–50.

Couch, J.A. 1967. "Concurrent haplosporidian infections of the oyster, *Crassostrea virginica* (Gmelin)." *J. Paras.* 53: 248–53.

Couch, J.A., Farley, C.A. and Rosenfield, A. 1966. "Sporulation of *Minchinia nelsoni* (Haplosporida, Haplosporidiidae) in *Crassostrea virginica* (Gmelin)." *Science* 153: 1529–31.

Coulson, J.C. 1971. "Competition for breeding sites causing segregation and reduced young production in colonial animals." In: den Boer, P.J. and Gradwell, G.R. eds.: *Dynamics of Populations. Proc. Adv. Study Inst. on*

Dynamics of Numbers in Populations. Oosterbeek, 1970. Centre for Agricultural Publ. Docum. Wageningen 257–68.

Cressey, R.F. and Lachner, E.A. 1970. "The parasitic copepod diet and life history of diskfishes (Echeneidae)." *Copeia* 1970. 2: 310–18.

Crofton, H.D. 1971. "A quantitative approach to parasitism." *Parasitol.* 62: 179–93.

Crompton, D.W.T. 1970. *An Ecological Approach to Acanthocephalan Physiology.* Cambridge: Cambridge University Press.

Crompton, D.W.T. 1973. "The sites occupied by some parasitic helminths in the alimentary tract of vertebrates." *Biol. Rev.* 48: 27–83.

Crompton, D.W.T. 1976. "Entry into the host and site selection." In: Kennedy C.R. ed.: *Ecological Aspects of Parasitology.* Amsterdam, Oxford: North Holland Publ. Co., 41–73.

Dailey, M.D. 1975. "Investigations on the viability of larval helminths after freezing." *Aquatic Mammals* 3: 22–25.

Dailey, M.D. and Brownell, R.L. jr. 1972. "A checklist of marine mammal parasites." In: Ridgway, S.H. ed.: *Mammals of the Sea. Biology and Medicine.* Springfield, Illinois, U.S.A.: Charles C. Thomas.

Dailey, M.D. and Ridgway, S.H. 1976. "A trematode from the round window of an Atlantic bottlenosed dolphin's ear." *J. Wildlife Diseases* 12: 45–47.

Dailey, M. and Stroud, R. 1978. "Parasites and associated pathology observed in cetaceans stranded along the Oregon Coast." *J. Wildlife Diseases* 14: 503–11.

Dailey, M.D. and Walker, W.A. 1978. "Parasitism as a factor (?) in single strandings of southern California cetaceans." *J. Paras.* 64: 593–96.

Davey, J.T. and Gee, J.M. 1976. "The occurrence of *Mytilicola intestinalis* Steuer, an intestinal copepod parasite of *Mytilus*, in the south-west of England." *J. mar. biol. Ass. U.K.* 56: 85–94.

Davis, D.S. and Farley, J. 1973. "The effect of parasitism by the trematode *Cryptocotyle lingua* (Creplin) on digestive efficiency in the snail host, *Littorina saxatilis* (Olivi)." *Parasitol.* 66: 191–97.

Dawes, B. 1930. "The histology of the alimentary tract of the plaice (*Pleuronectes platessa*)." *Quart. J. Micr. Sc.* 73: 243–74.

Dawes, B. 1964. *The Trematoda.* Cambridge: Cambridge University Press.

Dick, T.A. and Belosevic, M. 1978. "Parasites of Arctic char (*Salvelinus alpinus*) and their use in identifying marine (migratory) and freshwater (non-migratory) populations of char." In: *4th Inter. Congr. Paras.*, 19–26 August, 1978. Warsaw. Section H: 5.

Di Conza, J.J. and Halliday, W.J. 1971. "Relationship of catfish serum antibodies to immunoglobulin in mucus secretions." *Austr. J. exp. Biol. med. Sci.* 49: 517–19.

Dienske, H. 1968. "A survey of the metazoan parasites of the rabbit-fish, *Chimaera monstrosa* L. (Holocephali)." *Netherl. J. Sea Res.* 4: 32–58.

Dillon, W.A. 1966. "Provisional list of parasites occurring on *Fundulus* spp." *Virginia J. Sc.*, January 1966: 21–31.

Doflein, F. and Reichenow, E. 1953. *Lehrbuch der Protozoenkunde.* Jena: VEB Gustav Fischer Verlag.

Dogiel, V.A. 1964. *General Parasitology.* English translation Edinburgh–London: Oliver and Boyd.
Dogiel, V.A. 1965. *General Protozoology.* 2nd ed., English translation: Oxford: Clarendon Press.
Dogiel, V.A., Petrushevski, G.K. and Polyanski, Yu. I. 1961. *Parasitology of Fishes.* English translation Edinburgh and London: Oliver and Boyd.
Dollfus, R. Ph. 1946. "Parasites (animaux et végétaux) des helminthes. Hyperparasites, ennemis et prédateurs des helminthes parasites et des helminthes libres." *Encycl. biol.* 27: 1–483. Paris: Paul Lechevalier.
Dollfus, R.P. 1953. "Apercu général sur l'histoire naturelle des parasites animaux de la morue Atlanto–Arctique *Gadus callarias* L. (= *morhua* L.) et leur distribution géographique." *Encycl. Biol.* 43. Paris: Paul Lechevalier.
Dollfus, R. Ph. 1957. "Que savons-nous sur la spécificité parasitaire des cestodes Tétrarhynques?" *First Symp. Host Specificity Among Par. Vertebr.* Neuchâtel: Paul Attinger, 255–58.
Dubois, R. 1901. "Sur le mécanisme de la formation des perles fines dans le *Mytilus edulis.*" *Compt. Rend. Hebdom. Seances Acad. Sc.* 133: 603–5.
Eibl-Eibesfeldt, I. 1955. "Über Symbiosen, Parasitismus und andere besondere zwischenartliche Beziehungen tropischer Meeresfische." *Z. Tierpsych.* 12: 203–19.
Eibl-Eibesfeldt, I. 1959. "Der Fisch *Aspidontus taeniatus* als Nachahmer des Putzers *Labroides dimidiatus.*" *Z. Tierpsych.* 16: 19–25.
Eibl-Eibesfeldt, I. 1961. "Eine Symbiose zwischen Fischen (*Siphamia versicolor*) und Seeigeln." *Z. Tierpsych.* 18: 56–59.
Eichler, W. 1948. "Some rules in ectoparasitism." *Ann. Mag. nat. Hist.* 1: 588–98.
Ekman, S. 1953. *Zoogeography of the Sea.* London: Sidgwick and Jackson.
Elder, H.Y. 1979. "Studies on the host parasite relationship between the parasitic prosobranch *Thyca crystallina* and the asteroid starfish *Linckia laevigata.*" *J. Zool.* 187: 369–91.
Esch, G.W. 1977. *Regulation of Parasite Populations.* New York: Academic Press.
Euzet, L. 1957. "Cestodes de Sélachiens." *First Symp. Host Specificity among Par. Vertebr.* Neuchâtel: Paul Attinger, 259–69.
Ewers, W.H. 1961. "A new intermediate host of schistosome trematodes from New South Wales." *Nature* 190: 283–84.
Ewers, W.H. and Rose, C.R. 1965. "Trematode parasitism and polymorphism in a marine snail." *Science* 148: 1747–48.
Farley, C.A. 1967. "A proposed life cycle of *Minchinia nelsoni* (Haplosporida, Haplosporidiidae) in the American oyster *Crassostrea virginica.*" *J. Protozool.* 14: 616–25.
Farley, C.A. 1968. "*Minchinia nelsoni* (Haplosporida) disease syndrome in the American oyster *Crassostrea virginica.*" *J. Protozool.* 15: 585–99.
Feder, H.M. 1966. "Cleaning symbiosis in the marine environment." In: Henry, S.M. ed.: *Symbiosis.* London–New York: Acad. Press. 327–80.
Fell, H.B. 1961. "Ophiuroidea, Fauna of the Ross Sea." *Mem. N.Z. Oceanogr. Inst.* 18.

Fenchel, T. 1965. "Ciliates from Scandinavian molluscs." *Ophelia* 2: 71–174.
Fenchel, T. 1966. "On the ciliated Protozoa inhabiting the mantle cavity of lamellibranchs." *Malacologia* 5: 35–36.
Feng, S.Y. and Canzonier, W.J. 1970. "Humoral responses in the American oyster (*Crassostrea virginica*) infected with *Bucephalus* sp. and *Minchinia nelsoni*." In: Snieszko, S.F.: *A Symposium on Diseases of Fishes and Shellfishes*. *Amer. Fish. Soc.* Special Publ. 5, Washington, D.C., 497–510.
Fernando, C.H. and Hanek, C. 1976. "Gills." In: Kennedy, C.R. ed.: *Ecological Aspects of Parasitology*. Amsterdam: North-Holland Publ. Co.: 207–26.
Fiedler, K. 1964. "Verhaltensstudien an Lippfischen der Gattung *Crenilabrus* (Labridae, Perciformes)." *Z. Tierpsych.* 21: 521–91.
Fletcher, T.C. and Grant, P.T. 1969. "Immunoglobulins in the serum and mucus of the plaice (*Pleuronectes platessa*)." *Biochem. J.* 115: 1–65.
Freeman, R.F.H. 1958. "An adult digenetic trematode from an invertebrate host: *Proctoeces subtenuis* (Linton) from the lamellibranch *Scrobicularia plana* (de Costa)." *J. mar. biol. Ass. U.K.* 37: 435–57.
Fretter, V. and Graham, A. 1949. "The structure and mode of life on the Pyramidellidae, parasitic opisthobranchs." *J. mar. biol. Assoc. U.K.* 28: 493–532.
Fricke, H.W. 1966. "Zum Verhalten des Putzerfisches *Labroides dimidiatus*." *Z. Tierpsych.* 23: 1–3.
Fry, F.E.J. 1957. "The aquatic respiration of fish." In: Brown, M.E. ed.: *The Physiology of Fishes*, vol. 1., New York: Acad. Press, 1–63.
Geraci, J.R., Dailey, M.D. and St. Aubin, D.J. 1978. "Parasitic mastitis in the Atlantic white-sided dolphin, *Lagenorhynchus acutus*, as a probably factor in herd productivity." *J. Fisheries Res. Board Canada.* 35: 1350–55.
Ghiselin, M.T. 1969. "The evolution of hermaphroditism among animals." *Quart. Rev. Biol.* 44: 189–208.
Ghittino, P. 1974. "Present knowledge of the principal diseases of cultured marine fish." *Rivista Italiana di Piscicolture e Ittiopatologia.* 9: 51–56.
Giard, M.A. 1903. "Sur la production volontaire des perles fines on margarose artificielle." *Compt. Rend. Soc. Biol.*, Paris. 55: 1225–26.
Giard, M.A. 1907. "Sur les trématodes margaritigenès du Pas-du-Calais (*Gymnophallus somateriae* Levinsen et *G. bursicola* Odhner)." *Compt. Rend. Soc. Biol.*, Paris. 63: 416–20.
Gibson, R. 1967. "Occurrence of the entocommensal rhynchocoelan, *Malacobdella grossa*, in the oval piddock. *Zirfaea crispata*, on the Yorkshire coast." *J. mar. biol. Ass. U.K.* 47: 301–17.
Gibson, R. 1968. "Studies on the biology of the entocommensal rhynchocoelan *Malacobdella grossa*." *J. mar. biol. Ass. U.K.* 48: 637–56.
Gibson, D.I. 1972. "Flounder parasites as biological tags." *J. Fish. Biol.* 4: 1–9.
Gibson, R. and Jennings, J.B. 1969. "Observations on the diet, feeding mechanisms, digestion and food reserves of the entocommensal rhynochocoelan *Malacobdella grossa*." *J. mar. biol. Ass. U.K.* 49: 17–32.

Ginetzinskaya, T.A. and Dobrovolski, A.A. 1978. (*Special Parasitology*), vol. 1. Moscow: "Visschaja schkola" (in Russian).

Golvan, Y.J. 1957. "La Spécificité Parasitaire Chez les Acanthocéphales." *First. Symp. Host Specificity Among Par. Vertebr.* Neuchâtel. Paul Attinger, 244–54.

Gooding, R.M. 1964. "Observations of fish from a floating observation raft at sea." *Proc. Hawaiian Acad. Sci.* 39: 27.

Gooding, R.M. and Magnuson, J. 1964. Personal communication to Feder, H.M. (see Feder, H.M. 1966).

Gorlick, D.L., Atkins, P.D. and Losey G.S. jr. 1978. "Cleaning stations as water holes, garbage dumps, and sites for the evolution of reciprocal altruism." *Am. Natur.* 112: 341–53.

Gotshall, D.W. 1967. "Cleaning symbiosis in Monterey Bay, California." *Calif. Fish Game* 53: 125–26.

Grabda, E. 1976. "Ecological problems in fish parasitology." *Wiad. parazyt.* 22: 531–41.

Grabda, J. 1974. "The dynamics of the nematode larvae, *Anisakis simplex* (Rud.) invasion in the south-western Baltic herring (*Clupea harengus* L.)." *Acta Ichthy. Pisc.* 4: 3–21.

Grabda, J. 1976. "The occurrence of anisakid nematode larvae in Baltic cod (*Gadus morhua callarias* L.) and the dynamics of their invasion." *Acta Ichthy. et Pisc.* 6: 3–22.

Grabda, J. 1977a. "Studies on parasitisation and consumability of Alaska pollack, *Theragra chalcogramma* (Pall.)." *Acta Ichthy. et Pisc.* 7: 15–34.

Grabda, J. 1977b. (*An outline of parasitology of marine fishes.*) *Zarys morskiej parazytologii rybackiej.* Szczecin: Akademia Rolnicza, 190 pp. (In Polish).

Grabda, J. 1978. "Studies on parasitic infestation of blue whiting (*Micromesistius* spp.) with respect to the fish utilization for consumption." *Acta Ichthy. et Pisc.* 8: 29–41.

Grainger, J.N.R. 1951. "Notes on the biology of the copepod *Mytilicola intestinalis* Steuer." *Parasit.* 41: 135–42.

Grassé, P. ed. 1959. *Traité de Zoologie.* vol. V. Paris: Massou et Cie.

Grassé, P. ed. 1965. *Traité de Zoologie.* vol. IV. Paris: Massou et Cie.

Grell, K.G. 1956. *Protozoologie.* Berlin, Göttingen, Heidelberg: Springer.

Gress, F.M. and Cheng, T.C. 1973. "Alterations in total serum proteins and protein fractions in *Biomphalaria glabrata* parasitized by *Schistosoma mansoni.*" *J. Invert. Pathol.* 22: 382–90.

Gröben, G. 1940. "Beobachtungen über die Entwicklung verschiedener Arten von Fischschmarotzern aus der Gattung *Dactylogyrus.*" *Z. Parasitenk.* 11: 611–36.

Gusev, A.V. 1957. "(Parasitological investigations of some deep-sea fishes in the Pacific Ocean.)" *Trudy Inst. Oceanol.* 27: 362–66 (in Russian).

Gusev, A.V. and Fernando, C.H. 1973. "Dactylogyridae (Monogenoidea) from the stomach of fishes." *Folia Paras.* 20: 207–12.

Halton, D.W. 1974. "Hemoglobin absorption in the gut of a monogenetic trematode, *Diclidophora merlangi.*" *J. Paras.* 60: 59–66.

Halton, D.W. and Jennings, J.B. 1965. "Observations on the nutrition of monogenetic trematodes." *Biol. Bull.* 129: 257–72.
Halvörsen, O. 1976. "Negative interaction amongst parasites." In: Kennedy, C.R. ed.: *Ecological Aspects of Parasitology*. Amsterdam, Oxford: North Holland Publ. Co., 99–114.
Hamrick, T. 1974. "Whale beachings suicide? No, says Smithsonian team." *National Fisherman* 55: 12A, 23A.
Hanson, E.D. 1972. *Survival Diversity*. Englewood Cliffs, N.J.: Prentice-Hall.
Harant, H. 1931. "Les ascidies et leurs parasites." *Ann. Monaco Inst. Oceanogr.* 8: 230–389.
Hare, G.M. and Burt, M.D.B. 1976. "Parasites as potential biological tags of Atlantic salmon (*Salmo salar*) smolts in the Miramichi River System, New Brunswick." *J. Fish. Res. Board. Canada* 33: 1139–43.
Hargis, W.J. 1957. "The host specificity of monogenetic trematodes." *Exp. Paras.* 6: 610–25.
Hargis, W.J. 1958. "Parasites and fishery problems." *Proc. Gulf Carib. Fish Inst.* 1958, 70–75.
Hartnoll, R.G. 1960. "*Entionella monensis* sp. nov., an entoniscid parasite of the spider crab *Eurynome aspera* (Pennant)." *J. mar. biol. Ass. U.K.* 39: 101–107.
Haskin, H.H., Stauber, L.A. and Mackin, J.A. 1966. "*Minchinia nelsoni* n. sp. (Haplosporida, Haplosporidiidae) causative agent of the Delaware Bay oyster epizoötic." *Science* 153: 1414–16.
Haswell, W.A. 1886. "On a destructive parasite of the rock oyster." *Proc. Linn. Soc. N.S.W.* 10: 273–75.
Haswell, W.A. 1888. "On *Sacculina* infesting Australian crabs." *Proc. Linn. Soc. N.S.W.* 3: 1711–12.
Hauck, A.K. 1977. "Occurrence and survival of the larval nematode *Anisakis* sp. in the flesh of fresh, frozen, brined and smoked Pacific herring, *Clupea harengus pallasi*." *J. Paras.* 63: 515–19.
Haven, D. 1959. "Effects of pea crabs *Pinnotheres ostreum* on oysters *Crassostrea virginica*." *Proc. Nat. Shellfisheries Ass.* 49: 77–86.
Hayasaka, H., Ishikura, H. and Takayama, T. 1971. "Acute regional ileitis due to *Anisakis* larvae." *International Surgery* 55: 8–14.
Hedgpeth, J.W. 1957. "Marine biogeography." In: Hedgpeth, J.W. ed.: *Treatise on Marine Ecology and Paleoecology*. vol. 1. Geol. Soc. America: 359–82.
Hediger, H. 1953. "Ein symbioseartiges Verhältnis zwischen Flusspferd und Fisch." *Säugetierkdl. Mitt.* 1: 75–76.
Henrickson, R.C. 1967. "Incorporation of tritiated thymidine by teleost epidermal cells." *Experientia* 23: 357–58.
Hepper, B.T. 1955. "Environmental factors governing the infection of mussels, *Mytilus edulis*, by *Mytilicola intestinalis*." *Fisheries Invest., Min. Agric. Fish. Food.* ser. II. 20: 1–21.
Hepper, B.T. 1956. "The European flat oyster, *Ostrea edulis* L., as a host for *Mytilicola intestinalis* Steuer." *J. Anim. Ecol.* 25: 144–47.

Herald, E.S. 1964. "Cleanerfish for a cleaner aquarium." *Pacific Discovery* 17: 28–29.
Hespenheide, H.A. 1979. "Are there fewer parasitoids in the tropics?" *Am. Natur.* 113: 766–69.
Hipeau-Jacquotte, R. 1977. "Ethologie du stade infestant du copépode ascidicole Notodelphyidae *Pachypygus gibber*. I. Réactions à la lumière au cours de la vie larvaire pélagique." *Mar. Biol.* 44: 57–63.
Hislop, J.R.G. and MacKenzie, K. 1976. "Population studies of the Whiting *Merlangius merlangus* (L.) of the northern North Sea." *J. Conseil. Inter. l'Exp. Mer.* 37: 98–110.
Hobson, E.S. 1965. "Diurnal–nocturnal activity of some inshore fishes in the Gulf of Carpentaria." *Copeia* 1965. 3: 291–302.
Hobson, E.S. 1969. "Comments on certain recent generalizations regarding cleaning symbiosis in fishes." *Pacif. Sc.* 23: 35–39.
Hobson, E.S. 1971. "Cleaning symbiosis among California inshore fishes." *Fishery Bull.* 69: 491–523.
Hogue, M.J. 1921. "Studies on the life history of *Vahlkampfia patuxent* n. sp., parasitic in the oyster, with experiments regarding its pathogenicity." *Am J. Hyg.* 1: 321–45.
Holdich, D.M. and Harrison, K. 1980. "The crustacean isopod genus *Gnathia* Leach from Queensland waters with descriptions of nine new species." *Aust. J. Mar. Freshwater Res.* 31: 215–40.
Holmes, J.C. 1971. "Habitat segregation in sanguinicolid blood flukes (Digenea) of scorpaenid rockfishes (Perciformes) on the Pacific coast of North America." *J. Fisheries Res. Board. Canada* 28: 903–909.
Holmes, J.C. 1973. "Site selection by parasitic helminths: interspecific interactions, site segregation, and their importance to the development of helminth communities." *Can. J. Zool.* 51: 333–47.
Holmes, J.C. 1976. "Host selection and its consequences." In: Kennedy, C.R. ed.: *Ecological Aspects of Parasitology*. Amsterdam, Oxford: North Holland Publ. Co., 22–39.
Hooper, J. 1980. "The taxonomy and ecology of some parasites in marine flathead fishes (family Platycephalidae) from Northern New South Wales." M. Sc. thesis, University of New England, Armidale.
Hopkins, S.H. 1947. "The nemertean *Carcinonemertes* as an indicator of the spawning history of the host, *Callinectes sapidus*." *J. Paras.* 33: 146–50.
Hopkins, S.H. 1956a. "Notes on the boring sponges in Gulf coast estuaries and their relation to salinity." *Bull. mar. sc. of the Gulf and Caribbean* 6: 44–58.
Hopkins, S.H. 1956b. "The boring sponges which attack South Carolina oysters, with notes on some associated organisms." *Contrib. Bears Bluff. Lab.* 23: 1–30.
Hopkins, S.H. 1956c. "*Odostomia impressa* parasitizing southern oysters." *Science* 124: 628–29.
Hopkins, S.H. 1958. "Trematode parasites of *Donax variabilis* at Mustang Island, Texas." *Contr. Mar. Sc., Texas Univ., Mar. Sc. Inst.* 5: 301–311.
Hopkins, S.H. 1962. "Distribution of species of *Cliona* (boring sponge) on

the eastern shore of Virginia in relation to salinity." *Chesapeake Science.* 3: 121–24.
Horsley, R.A. 1977. "A review of the bacterial flora of teleosts and elasmobranchs, including methods for its analysis." *J. Fish. Biol.* 10: 529–53.
Hotta, H. 1962. "The parasitism of saury (*Cololabis saura*) infected with the parasitic copepod *Caligus macarovi* Gussev, during the fishing season in 1961." *Bull. Tohoku Reg. Fish. Res. Lab.* 21: 50–56, (cited by Kabata, 1970).
Huff, C.G. and Shiroishi, T. 1962. "Natural infection of Humboldt's penguin with *Plasmodium elongatum*." *J. Paras.* 48: 495.
Hughes, G.M. 1960. "A comparative study of gill ventilation in marine teleosts." *J. Exp. Biol.* 37: 28–45.
Hughes, G.M. and Grimstone, A.V. 1965. "The fine structure of the secondary lamellae of the gills of *Gadus pollachius*." *Quart. J. micros. Sc.* 106: 343–53.
Hughes, G.M. and Iwai, T. 1978. "A morphometric study of the gills in some Pacific deep-sea fishes." *J. Zool.* 184: 155–70.
Humes, A.G. 1942. "The morphology, taxonomy, and bionomics of the nemertean genus *Carcinonemertes*." *Ill. Biol. Monogr.* 18: 1–105.
Humes, A.G. 1954. "*Mytilicola porrecta* n. sp. (Copepoda: Cyclopoida) from the intestine of marine pelecypods." *J. Paras.* 40: 186–94.
Hunter, C.R. and Nayudu, P.L. 1978. "Surface folds in superficial epidermal cells of three species of teleost fish." *J. Fish. Biol.* 12: 163–66.
Hutton, R.F. 1964. "A second list of parasites from marine and coastal animals of Florida." *Trans. Amer. Microsc. Soc.* 79: 287–92.
Hutton, R.F. and Sogandares-Bernal, F. 1960. "A list of parasites from marine and coastal animals of Florida." *Trans. Amer. Microsc. Soc.* 79: 287–92.
Hyman, L.H. 1951. *The Invertebrates.* Vol. 2. Platyhelminthes *and* Rhynchocoela. New York: McGraw-Hill.
Ihering, H. von. 1891. "On the ancient relations between New Zealand and South America." *Trans. Proc. New Zealand Inst.* 24: 431–45.
Ihering, H. von. 1902. "Die Helminthen als Hilfsmittel der zoogeographischen Forschung." *Zool. Anz.* 26: 24–51.
Ivanov, V.P. and Markov, G.S. 1973. "(Helminths of sturgeon hybrids (USSR).)" *Materialy Nauc. Konf. Vsesoy. Obsh. Gel'mint.* No. 25: 122–24 (In Russian).
Izjumova, N.A. 1953. "(Biology of *Dactylogyrus vastator* Nybelin and *Dactylogyrus solidus* Achmerov in Carp farms.)" *Zool. Inst. AN SSSR, autoref diss.* 1–17 (in Russian).
Jahn, T.L. and Kuhn, L.R. 1932. "The life history of *Epibdella melleni* MacCallum 1927, a monogenetic trematode parasitic on marine fishes." *Biol. Bull.* 62: 89–111.
James, B.L. 1965. "The effects of parasitism by larval Digenea on the digestive gland of the intertidal prosobranch, *Littorina saxatilis* (Olivi) subsp. *tenebrosa* (Montagu)." *Parasitol.* 55: 93–115.

James, B.L. 1971. "Host selection and ecology of marine digenean larvae." In: Crisp, D.J. ed.: *Fourth European Marine Biology Symposium.* Cambridge: Cambridge Univ. Press. 179–96.

James, B.L. and Bowers, E.A. 1967a. "The effects of parasitism by the daughter sporocyst of *Cercaria bucephalopsis haimaena* Lacaze-Duthiers, 1854, on the digestive tubules of the cockle, *Cardium edule L.*" *Parasitol.* 57: 67–77.

James, B.L. and Bowers, E.A. 1967b. "Histochemical observations on the occurrence of carbohydrates, lipids and enzymes in the daughter sporocyst of *Cercaria bucephalopsis haimaena* Lacaze-Duthiers, 1854 (Digenea: Bucephalidae)." *Parasitol.* 57: 79–86.

Jameson, H.L. 1912. "Studies on pearl oysters and pearls I. The structure of the shell and pearls of the Ceylon pearl-oyster (*Margaritifera vulgaris* Schumacher): with an examination of the cestode theory of pearl-formation." *Proc. Zool. Soc. London,* 1912: 260–358.

Jennings, J.B. and Calow, P. 1975. "The relationship between high fecundity and the evolution of entoparasitism." *Oecologia* 21: 109–115.

Jennings, J.B. and Gelder, S.R. 1976. "Observations on the feeding mechanism, diet and digestive physiology of *Histriobdella homari* van Beneden 1858: an aberrant polychaete symbiotic with North American and European lobsters." *Biol. Bull.* 151: 489–517.

Jepps, M.W. 1937. "On the protozoan parasites of *Calanus finmarchicus* in the Clyde sea area." *Quart. J. Micr. Sc.* 79, new ser. 589–658.

Jetté, E. 1977. "Helminth endoparasites of Sweet Lip and Coral Trout." M. Sc. thesis, James Cook University of North Queensland.

Johnston, T.H. 1952. "List of titles of the published works of Thomas Harvey Johnston, M.A., D. Sc." *Trans. Roy. Soc. S. Austr.* 75: I–IX.

Johnston, T.H. and Cleland, J.B. 1910. "The haematozoa of Australian fish. I." *Proc. Roy. Soc. N.S.W.* 44: 406–415.

Juranek, D.D. and Schultz, M.G. 1978. "Trichinellosis in humans in the United States: epidemiologic trends." In: Kim, C.W. and Pawlowski, Z.S. ed.: *Trichinellosis. Proc. 4th Inter. Conf. Trichinellosis.* Hanover, New Hampshire: University Press of New England.

Kabata, Z. 1970. "Diseases of fishes. I." *Crustacea as Enemies of Fishes.* Jersey City: T.F.H. Publ.

Kabata, Z. 1979. *Parasitic Copepoda of British Fishes.* London: Roy. Soc.

Kaestner, A. 1954–1963. *Lehrbuch der Speziellen Zoologie.* Jena: VEB Gustav Fischer Verlag.

Kagei, N., Sano, M., Takahashi, Y., Tamura, Y. and Sakamoto, M. 1978. "A case of acute abdominal syndrome caused by *Anisakis* type-II larva." *Jap. J. Parasit.* 27: 427–31.

Kanneworff, B. and Christensen, A.M. 1966. "*Kronborgia caridicola* sp. nov., an endoparasitic turbellarian from North American shrimps." *Ophelia* 3: 65–80.

Kapoor, B.G., Smit, H. and Verighina, I.A. 1975. "The alimentary canal and digestion in teleosts." *Adv. mar. Biol.* 13: 109–239.

Katkansky, S.C., Sparks, A.K. and Chew, K.K. 1967. "Distribution and effects of the endoparasitic copepod, *Mytilicola orientalis*, on the Pacific oyster, *Crassostrea gigas*, on the Pacific coast." *Proc. Nat. Shellfish. Assoc.* 57: 50–58.

Kearn, G.C. 1962. "Breathing movements in *Entobdella soleae* (Trematoda, Monogenea) from the skin of the common sole." *J. mar. biol. Ass. U.K.* 42: 93–104.

Kearn, G.C. 1963a. "The life cycle of the monogenean *Entobdella soleae*, a skin parasite of the common sole." *Parasitol.* 53: 253–63.

Kearn, G.C. 1963b. "Feeding in some monogenean skin parasites: *Entobdella soleae* on *Solea solea* and *Acanthocotyle* sp. on *Raja clavata*." *J. mar. biol. Ass. U.K.* 43: 749–66.

Kearn, G.C. 1967. "Experiments on host-finding and host-specificity in the monogenean skin parasite *Entobdella soleae*." *Parasitol.* 57: 585–605.

Kearn, G.C. 1970a. "The production, transfer and assimilation of spermatophores by *Entobdella soleae*, a monogenean skin parasite of the common sole." *Parasitol.* 60: 301–311.

Kearn, G.C. 1970b. "The physiology and behaviour of the monogenean skin parasite *Entobdella soleae* in relation to its host (*Solea solea*)." In: Fallis, A.M. ed.: *Ecology and Physiology of Parasites*. London: Adam Hilger, 161–87.

Kearn, G.C. 1971. "The attachment site, invasion route and larval development of *Trochopus pini*, a monogenean from the gills of *Trigla hirundo*." *Parasitol.* 63: 513–25.

Kearn, G.C. 1973. "An endogenous circadian hatching rhythm in the monogenean skin parasite *Entobdella soleae*, and its relationship to the activity of the host (*Solea solea*)." *Parasitol.* 66: 101–22.

Kearn, G.C. 1974a. "Nocturnal hatching in the monogenean skin parasite *Entobdella hippoglossi* from the halibut, *Hippoglossus hippoglossus*." *Parasitol.* 68: 161–72.

Kearn, G.C. 1974b. "The effects of fish skin mucus on hatching in the monogenean parasite *Entobdella soleae* from the skin of the common sole, *Solea solea*." *Parasitol.* 68: 173–88.

Kearn, G.C. 1974c. "A comparative study of the glandular and excretory systems of the oncomiracidia of the monogenean skin parasites *Entobdella hippoglossi*, *E. diadema* and *E. soleae*." *Parasitol.* 69: 257–69.

Kearn, G.C. 1975a. "Hatching in the monogenean parasite *Dictyocotyle coeliaca* from the body cavity of *Raja naevus*." *Parasitol.* 70: 87–93.

Kearn, G.C. 1975b. "The mode of hatching of the monogenean *Entobdella soleae*, a skin parasite of the common sole (*Solea solea*)." *Parasitol.* 71: 419–31.

Kearn, G.C. 1976. "Body surface of fishes." In: Kennedy, C.R. ed.: *Ecological Aspects of Parasitology*. Amsterdam: North-Holland Publ. Co., 185–208.

Kearn, G.C. 1978. "Predation on a skin-parasitic monogenean by a fish." *J. Paras.* 64: 1129–30.

Kearn, G.C. and MacDonald, S. 1976. "The chemical nature of the host hatching factors in monogenean skin parasites *Entobdella soleae* and *Acanthocotyle lobianchi*." *Internat. J. Paras.* 6: 457–66.

Kennedy, C.R. 1975. *Ecological Animal Parasitology.* Oxford, London, Edinburgh, Melbourne: Blackwell Sc. Publ.
Kennedy, C.R. ed. 1976a. *Ecological Aspects of Parasitology.* Amsterdam, Oxford: North Holland Publ. Co.
Kennedy, C.R. 1976b. "Reproduction and dispersal." In: Kennedy, C.R. ed.: *Ecological Aspects of Parasitology.* Amsterdam, Oxford: North Holland Publ. Co., 143–60.
Kennedy, C.R. 1977. "The regulation of fish parasite populations." In: Esch, G.W. ed.: *Regulations of Parasite Populations.* New York–London–San Francisco: Acad. Press.
Kennedy, C.R. 1979. "The distribution and biology of the cestode *Eubothrium parvum* in capelin, *Mallotus villosus* (Pallas) in the Barents Sea, and its use as a biological tag." *J. Fish. Biol.* 15: 223–36.
Key, D. and Alderman, D.J. 1974. "A bibliography of diseases of commercial bivalves (supplementary to Sindermann 1968), with annotations upon recent European work." *Fisheries Lab. Tech. Rep. Series. U.K.* No. 11. 35 pp.
Khalil, L.F. 1964. "On the biology of *Macrogyrodactylus polyptiri* Malmberg, 1956, a monogenetic trematode on *Polypterus senegalus* in the Sudan." *J. Helminth.* 38: 219–22.
Khalil, L.F. 1970. "Further studies on *Macrogyrodactylus polypteri*, a monogenean on the African freshwater fish *Polypterus senegalus*." *J. Helminth.* 44: 329–48.
Khan, R.A. 1976. "The life cycle of *Trypanosoma murmanensis* Nikitin." *Can. J. Zool.* 54: 1840–49.
Kikkawa, J. 1977. "Ecological paradoxes." *Aust. J. Ecol.* 2: 121–36.
Kirmse, P.D. 1978. "*Haemogregarina sachai* n. sp. from cultured turbot *Scophthalmus maximus* (L.) in Scotland." *J. Fish Dis.* 1: 337–42.
Ko, R.C. 1975. "*Echinocephalus sinensis* n. sp. (Nematoda: Gnathostomatidae) from the ray (*Aetabutus flagellum*) in Hong Kong, Southern China." *Can. J. Zool.* 53: 490–500.
Ko, R.C. 1976. "Experimental infection of mammals with larval *Echinocephalus sinensis* (Nematoda: Gnathostomatidae) from oysters (*Crassostrea gigas*)." *Can. J. Zool.* 54: 597–609.
Ko, R.C. 1977. "Effects of temperature acclimation on infection of *Echinocephalus sinensis* (Nematoda: Gnathostomatidae) from oysters to kittens." *Can. J. Zool.* 55: 1129–32.
Ko, R.C., Morton, B. and Wong, P.S. 1974. "*Echinocephalus* sp. Molin, 1858 (Spiruroidea: Gnathostomatidae), an unusual nematode from the oyster, *Crassostrea gigas* Thunberg, 1793." In: *3rd Inter. Congr. Paras. Munich*, Aug. 25–31. Proc. Vol. 3. FACTA publ. pp. 1731–32.
Ko, R.C., Morton, B. and Wong, P.S. 1975. "Prevalence and histopathology of *Echinocephalus sinensis* (Nematoda: Gnathostomatidae) in natural and experimental hosts." *Can. J. Zool.* 53: 550–59.
Kohlmann, F.W. 1961. "Untersuchungen zur Biologie, Anatomie und Histologie von *Polystoma integerrimum* Fröhlich." *Z. Parasitenk.* 20: 495–524.

Kollatsch, D. 1959. "Untersuchungen über die Biologie und Ökologie der Karpfenlaus (*Argulus foliaceus* L.)." *Zool. Beitr.* 5: 1–36.
Komai, T. 1922. *Studies on Two Aberrant Ctenophores, Coeloplana* and *Gastrodes.* Publ. by author. Kyoto. 102 pp.
Korringa, P. 1952a. "Epidemiological observations on the mussel parasite *Mytilicola intestinalis* Steuer, carried out in the Netherlands 1952." *Ann. Biol.* 9: 219–24.
Korringa, P. 1952b. "Recent advances in oyster biology." *Quart. Rev. Biol.* 27: 266–308, 339–65.
Korringa, P. 1954. "The shell of *Ostrea edulis* as a habitat." *Arch. Neerland. Zool.* 10: 32–152.
Korringa, P. 1957. "Epidemiological observations on the mussel parasite *Mytilicola intestinalis* Steuer, undertaken in the Netherlands in 1955." *Ann. Biol.*, Copenhagen 12: 230–31.
Korringa, P. 1968. "On the ecology and distribution of the parasitic copepod *Mytilicola intestinalis* Steuer." *Bijdragen tot de Dierkunde* 38: 47–57.
Kozlov, D.P. 1971. "(The ways in which pinnipeds become infected with *Trichinella*.)" *Trudy gel'mint. Lab.* 21: 36–40 (in Russian).
Kozlov, D.P. and Berezantsev, Yu.A. 1968. "(Occurrence of *Trichinella* in a walrus of the USSR.)" *Trudy gel'mint. Lab.* 19: 86–89 (in Russian).
Ktari, M.H. 1971. *Recherches sur la reproduction et le développement de quelques monogenes (Polyopisthocotylea) parasites de poissons marins.* Thèse Academic de Montpellier, Univ. Sc. Techn. Languedoc.
Kuris, A.M. 1974. "Trophic interactions: similarity of parasitic castrators to parasitoids." *Quart. Rev. Biol.* 49: 129–48.
Lackie, A.M. 1975. "The activation of infective stages of endoparasites of vertebrates." *Biol. Rev.* 50: 285–323.
Lahav, M. 1974. "The occurrence and control of parasites infecting Mugilidae in fish ponds in Israel." *Bamidgeh* 26: 99–103.
Laird, M. 1950. "*Henneguya vitiensis* n. sp., a myxosporidian from a Fijian marine fish, *Leiognathus fasciatus* (Lacépède, 1803)." *J. Paras.* 36: 285–92.
Laird, M. 1951a. "Studies on the trypanosomes of New Zealand fish." *Proc. Zool. Soc. London.* 121: 285–309.
Laird, M. 1951b. "A contribution to the study of Fijian haematozoa, with descriptions of a new species from each of the genera *Haemogregarina* and *Macrofilaria.*" *Zool. Publ. Victoria Uni.* No. 10.
Laird, M. 1952. "New haemogregarines from New Zealand marine fishes." *Trans. Roy. Soc. New Zealand* 79: 589–600.
Laird, M. 1953a. "The protozoa of New Zealand intertidal zone fishes." *Trans. Roy. Soc. New Zealand* 81: 79–143.
Laird, M. 1953b. "Parasites of South Pacific fishes I. Introduction, and haematozoa." *Can. J. Zool.* 36: 153–65.
Laird, M. 1961. "Microecological factors in oyster epizootics." *Can J. Zool.* 39: 449–85.
Laird, M. 1969. "Marine fish haematozoa from New Brunswick and New England." *J. Fisheries Res. Bd. Canada* 26: 1075–1102.

Lambert, S. and Maillard, C. 1975. "Repartition branchiale de deux monogenes: *Diplectanum aequans* (Wagener, 1857) Diesing, 1858 et *D. laubieri* Lambert and Maillard, 1974 (Monogenea: Monopisthocotylea) parasites simultanes de *Dicentrarchus labrax* (Teleosteen)." *Ann. Paras. Hum. Comp.* 50: 691–99.

Larrañeta, M.G. 1957. "Presence du parasite *Peroderma cylindricum* Heller sur la sardine de Castellon et d'Alicante." *Conseil Gen. Pech. Medit., Débats Doc. Tech.*, no. 4. 109–12.

Lauckner, G. 1971. "Zur Trematodenfauna der Herzmuscheln *Cardium edule* und *Cardium lamarcki.*" *Helgoländer Wissenschaftliche Meeresuntersuchungen.* 22: 277–400.

Laurie, J.S. 1971. "Carbohydrate absorption by *Gyrocotyle fimbriata* and *Gyrocotyle parvispinosa* (Platyhelminthes)." *Exp. Paras.* 29: 375–85.

Leader, J.P. and Bedford, J.J. 1979. "Oviposition by the marine caddis-fly, *Philanisus plebeius* (Walk.)." *Search* 10: 275–76.

Lebedev, B.I. 1969a. "(Discorrelative–symmetrical heterotopies of the organs in *Pentatres sphyraenae* (Monogenoidea).)" *Parazitologiya* 3: 149–57 (in Russian).

Lebedev, B.I. 1969a. "(Basic regularities in the distribution of monogeneans and trematodes of marine fishes in the world ocean.)" *Zool. J.* 48: 41–50 (in Russian).

Lee, D.L. 1971. "Helminthes as vectors of micro-organisms." In: Fallis, A.M. ed.: *Ecology and Physiology of Parasites.* London: Adam Hilger.

Léger, L. and Hollande, A.C. 1917. "Sur un nouveau protiste a facies de *Chytridiopsis*, parasite des ovules de l'Huitre." *Compt. Rend. Séances Soc. Biol.* 80: 61–64.

Leigh-Sharpe, W.H. 1935/36. "A list of British invertebrates with their characteristic parasitic and commensal Copepoda." *J. mar. biol. Assoc. U.K.* 20: 47–48.

Lester, R.J.G. 1972. "Attachment of *Gyrodactylus* to *Gasterosteus* and host response." *J. Paras.* 58: 717–22.

Lester, R.J.G. and Adams, J.R. 1974. "*Gyrodactylus alexanderi*: reproduction, mortality, and effect on its host *Gasterosteus aculeatus.*" *Can. J. Zool.* 52: 827–33.

Levine, N.D., Corliss, J.O., Cox, F.E.G., Deroux, G., Grain, J., Honigberg, B.M., Leedale, G.F., Loeblich, III., A.R., Lom, J., Lynn, D., Merinfold, E.G., Page, F.C., Poljansky, G., Sprague, V., Vavra, J. and Wallace, F.G. 1980. "A newly revised classification of the Protozoa." *J. Protozool.* 27: 37–58.

Li, S.Y. and Hsu, H.F. 1951. "On the frequency distribution of helminths in their naturally infected hosts." *J. Paras.* 37: 32–41.

Lichtenfels, J.R., Bier, J.W. and Madden, P.A. 1978. "Larval anisakid (*Sulcascaris*) nematodes from Atlantic molluscs with marine turtles as definite hosts." *Trans. Am. Microsc. Soc.* 97: 199–207.

Lie, K.J., Basch, P.F. and Heyneman, D. 1968. "Direct and indirect antagonism between *Paryphostomum segregatum* and *Echinostoma paraensei* in the snail *Biomphalaria glabrata.*" *Z. Parasitenk.* 31: 101–107.

Limbaugh, C. 1961. "Cleaning symbiosis." *Sci. Amer.* 205: 42–49.
Limbaugh, C., Pederson, H. and Chase, jr. F.A. 1961. "Shrimps that clean fishes." *Bull. Mar. Sc.* 11: 237–57.
Lincicome, D.R. 1963. "Chemical basis of parasitism." *Ann. N.Y. Acad. Sci.* 113: 360–80.
Lincicome, D.R. 1971. "The goodness of parasitism: a new hypothesis." In: Cheng, T.C. ed.: *Aspects of the Biology of Symbiosis.* Baltimore, Butterworth, London: University Park Press.
Lipin, A. 1911. "Die Morphologie und Biologie von *Polypodium hydriforme* Uss." *Zool. Jb. Anat. Ontog.* 31: 317–426.
Llewellyn, J. 1954. "Observations on the food and gut pigment of the *Polyopisthocotylea* (Trematoda: Monogenea)." *Parasitol.* 44: 428–37.
Llewellyn, J. 1956. "The host-specificity, micro-ecology, adhesive attitudes, and comparative morphology of some trematode gill parasites." *J. mar. biol. Assoc. U.K.* 35: 113–27.
Llewellyn, J. 1957. "Host-specificity in monogenetic trematodes." *First Symp. Host Specificity Among Par. Vertebr.* Neuchâtel: Paul Attinger, 199–212.
Llewellyn, J. 1959. "The larval development of two species of gastrocotylid trematode parasites from the gills of *Trachurus trachurus.*" *J. mar. biol. Assoc. U.K.* 38: 461–67.
Llewellyn, J. 1962. "The life histories and population dynamics of monogenean gill parasites of *Trachurus trachurus* (L.)." *J. mar. biol. Ass. U.K.* 42: 587–600.
Llewellyn, J. 1966. "The effects of fish hosts upon the body shapes of their monogenean parasites." *Proc. First Int. Congr. Parasit.* 1: 543–45.
Llewellyn, L.C. 1965. "Some aspects of the biology of the marine leech *Hemibdella soleae.*" *Proc. Zool. Soc. Lond.* 145: 509–28.
Lom, J. 1970. "Protozoa causing diseases in marine fishes." In: Snieszko, S.F. ed.: *A Symposium on Diseases of Fishes and Shellfishes.* Amer. Fish. Soc., Special Publ. 5, Washington, D.C., 101–23.
Lom, J. and Laird, M. 1969. "Parasitic protozoa from marine and euryhaline fish of Newfoundland and New Brunswick. I. Peritrichous ciliates." *Can. J. Zool.* 47: 1367–80.
Losanoff, V.L. 1956. "Two obscure oyster enemies in New England waters." *Science* 123: 1119–20.
Losey, G.S. jr. 1971. "Communication between fishes in cleaning symbiosis." In: Cheng, T.C. ed.: *Aspects of the Biology of Symbiosis.* Baltimore. University Park Press. 45–76.
Losey, G.S. 1972a. "The ecological importance of cleaning symbiosis." *Copeia* 1972: 820–33.
Losey, G.S. jr. 1972b. "Behavioral ecology of the cleaning fish." *Aust. Nat. Hist. Sept.* 1972: 232–38.
Losey, G.S. jr. 1975. "Cleaning symbiosis and other allogrooming behaviour. *Abst. Animal Behav. Soc.* 1975: 46.
Losey, G.S. jr. 1976. "Animal models: do they fool them or us?" *Abst. Animal Behav. Soc.* 1976: 41.

Losey, G.S. jr. 1977. "The validity of animal models: a test for cleaning symbiosis." *Biol. of Behaviour* (cited by Thresher, 1977).
Losey, G.S. jr. 1979. "Fish cleaning symbiosis: proximate causes of host behaviour." *Anim. Beh.* 27: 669–85.
Losey, G.S. jr. and Margules, L. 1974. "Cleaning symbiosis provides a positive reinforcer for fish." *Science* 184: 179–80.
Lüling, K.H. 1953. "Gewebeschäden durch parasitäre Copepoden besonders durch *Elytrophora bachyptera*." *Z. Parasitenk.* 16: 84–92.
Lunz, G.R. jr. 1940. "The annelid worm, *Polydora*, as an oyster pest." *Science* 92: 310.
Lunz, G.R. jr. 1941. "*Polydora*, a pest in South Carolina oysters." *J. Elisha Mitchell Sc. Soc.* 57: 273–83.
Lyons, K.M. 1973. "The epidermis and sense organs of the Monogenea and some related groups." *Adv. Paras.* 11: 193–232.
Lysagth, A.M. 1941. "The biology and trematode parasites of the gastropod *Littorina neritoides* (L.) on the Plymouth breakwater." *J. mar. biol. Assoc. U.K.* 25: 41–67.
MacDonald, S. 1974. "Host skin mucus as a hatching stimulant in *Acanthocotyle lobianchi*, a monogenean from the skin of *Raja* spp." *Parasitol.* 68: 331–38.
MacDonald, S. 1975. "Hatching rhythms in three species of *Diclidophora* (Monogenea) with observations on host behaviour." *Parasitol.* 71: 211–28.
MacInnis, A.J. 1976. "How parasites find hosts: some thoughts on the inception of host–parasite integration." In: Kennedy, C.R. ed.: *Ecological Aspects of Parasitology*. Amsterdam, Oxford: North Holland Publ. Co., 3–20.
MacKenzie, K. 1978. "Parasites as biological tags for herring, *Clupea harengus* L." In: *4th Inter. Congr. Paras.*, 19–26 August, 1978. Warsaw. Section H: 6.
MacKenzie, K. and Gibson, D. 1970. "Ecological studies of some parasites of plaice, *Pleuronectes platessa* (L.) and flounder, *Platichthys flesus* (L.). In: Taylor, A.E.R. and Muller, R. ed.: *Aspects of Fish Parasitology*. vol. 8. Oxford and Edinburgh: Blackwell Publ.
Mackerras, I.M. and Mackerras, M.J. 1925. "The haematozoa of Australian marine teleostei." *Proc. Linn. Soc. N.S.W.* Sydney 50: 359–66.
Mackin, J.G. 1961. "Mortalities of oysters." *Proc. Nat. Shellfish Ass.* 51: 21–40.
McLaren, D.J. 1976. "Nematode sense organs." *Adv. Paras.* 14: 195–265.
MacNeill, A.C., Neufeld, J.L. and Webster, W.A. 1975. "Pulmonary nematodiasis in a narwhal." *Can. Vet. J.* 16: 53–55.
McCutcheon, F.H. and McCutcheon, A.E. 1964. "Symbiotic behaviour among fishes from temperate ocean waters." *Science* 145: 948–49.
McVicar, A.H. 1972. "The ultrastructure of the parasite–host interface of three tetraphyllidean tapeworms of the elasmobranch *Raja naevus*." *Parasitol.* 65: 77–88.
McVicar, A.H. and Fletcher, T.C. 1970. "Serum factors in *Raja radiata* toxic

to *Acanthobothrium quadripartitum* (Cestoda: Tetraphyllidea), a parasite specific to *R. naevus.*" *Parasitol.* 61: 55–63.
McVicar, A.H. and MacKenzie, K. 1974. "Parasitological problems in marine fish farming." In: *3rd Inter. Congr. Paras. Munich.* August 25–31. Proc. Vol. 3. FACTA Publ. pp. 1719–20.
McVicar, A.H. and MacKenzie, K. 1977. "Effects of different systems of monoculture on marine fish parasites." In: Cherrett, J.M. and Sagar, G.R. ed.: *Origins of Pest, Parasite, Disease and Weed Problems.* Oxford. U.K., Blackwell Sc. Publ., 163–82.
Mann, H. 1951. "Qualitätsverminderung des Fleisches der Miesmuscheln durch den Befall mit *Mytilicola intestinalis.*" *Fischereiwelt* 3: 121–22.
Mann, H. 1954. "Die wirtschaftliche Bedeutung von Krankheiten bei Seefischen." *Die Fischwirtschaft* 8: 38–39.
Mann, H. 1970. "Copepoda and Isopoda as parasites of marine fishes." In: Snieszko, S.F. ed.: *A Symposium on Diseases of Fishes and Shellfishes.* Amer. Fish. Soc., Special Publ. 5, Washington, D.C., 177–89.
Mann, J.A. 1978. "Diseases and parasites of fishes: an annotated bibliography of books and symposia, 1904–1977." *Fish Disease Leaflet, Fish and Wildlife Service, United States Department of the Interior.* No. 53, 28 pp.
Manter, H.W. 1934. "Some digenetic trematodes from deep-water fish at Tortugas, Florida." *Carnegie Instit. Publ. no. 435. Papers from Tortugas Lab.* 27: 257–345.
Manter, H.W. 1940a. "Digenetic trematodes of fishes from Galapagos Islands and neighbouring Pacific." *Rept. Allan Hancock Pacif. Exped.* 2: 11–22.
Manter, H.W. 1940b. "The geographical distribution of digenetic trematodes of marine fishes of the tropical American Pacific." *Rept. Allan Hancock Pacific Exped.* 2: 531–47.
Manter, H.W. 1947. "The digenetic trematodes of marine fishes of Tortugas, Florida." *Am. Midland. Nat.* 38: 257–416.
Manter, H.W. 1951. "Studies on *Gyrocotyle rugosa* Diesing, 1850, a cestodarian parasite of the elephant fish, *Callorhynchus milii.*" *Zool. Publ. Victoria College.* 17: 1–11.
Manter, H.W. 1954a. "Some digenetic trematodes from fishes of New Zealand." *Trans. Roy. Soc. New Zealand* 82: 475–568.
Manter, H.W. 1954b. "Trematoda of the Gulf of Mexico." *Fish. Bull. Fish and Wildlife Serv.* 55: 335–50.
Manter, H.W. 1955. "The zoogeography of trematodes of marine fishes." *Exp. Paras.* 4: 62–86.
Manter, H.W. 1957. "Host specificity and other host relationships among the digenetic trematodes of marine fishes." *First Symp. Host Specificity Among Par. Vertebr.* Neuchâtel; Paul Attinger, 185–98.
Manter, H.W. 1966. "Parasites of fishes as biological indicators of recent and ancient conditions." In: McCauley, J.E. ed.: *Host Parasite Relationships.* Proc. 26th Ann. Biol. Colloq. Oregon State Univ. Press.
Manter, H.W. 1967. "Some aspects of the geographical distribution of parasites." *J. Paras.* 53: 1–9.

Margolis, L. 1970. "Nematode diseases of marine fishes." In: Snieszko, S.F. ed.: *A Symposium on Diseases of Fishes and Shellfishes.* Amer. Fish Soc., Special Publ. 5, Washington, D.C., 190–208.

Margolis, L. 1977. "Public health aspects of 'codworm' infection: a review." *J. Fish. Res. Bd. Can.* 34: 887–98.

Martin, W.E. 1969. "*Rynkatorpa pawsoni* n. sp. (Echinodermata: Holothuroidea) a commensal sea cucumber." *Biol. Bull.* 137: 332–37.

Martin, R. and Brinckmann, A. 1963. "Zum Brutparasitismus von *Phyllirrhoe bucephala* Per. and Les. (Gastropoda, Nudibranchia) auf der Meduse *Zanclea costata* Gegenb. (Hydrozoa, Anthomedusae)." *Pubbl. staz. zool. Napoli.* 33: 206–223.

Matthews, R.A. 1974. "The life-cycle of *Bucephaloides gracilescens* (Rudolphi, 1819) Hopkins, 1954 (Digenea: Gasterostomata)." *Parasitol.* 68: 1–12.

Maul, G.E. 1956. "Monografia dos peixes do Museu Municipal do Funchal. Ordem Discocephali." *Bol. Museu Munic. Funchal.* 9: 5–75.

May, R.M. 1977. "Dynamical aspects of host–parasite associations: Crofton's model revisited." *Parasitol.* 75: 259–76.

May, R.M. and Anderson, R.M. 1979. "Population biology of infectious diseases: Part II." *Nature* 280: 455–61.

Mayr, E., Linsley, E.G. and Usinger, R.L. 1953. *Methods and Principles of Systematic Zoology.* New York: McGraw-Hill.

Menzel, R.W. and Hopkins, S.H. 1955. "The growth of oysters parasitized by the fungus *Dermocystidium marinum* and by trematode *Bucephalus cuculus*." *J. Paras.* 41: 333–42.

Merrill, A.S. 1967. "Shell deformity of mollusks attributable to the hydroid, *Hydractinia echinata*." *Fisheries Bull.* 66: 273–79.

Meserve, T.G. 1938. "Some monogenetic trematodes from the Galapagos Islands and the neighbouring Pacific." *Rept. Allan Hancock Pacif. Exp.* 2: 31–89.

Metcalf, M.M. 1929. "Parasites and the aid they give in problems of taxonomy, geographical distribution, and paleogeography." *Smiths. Misc. Coll.* 81: 1–81.

Van Meter, V.B. and Ache, B.W. 1974. "Host location by the pearlfish *Carapus bermudensis*." *Mar. Biol.* 26: 379–83.

Meyer, P.F. and Mann, H. 1950. "Beiträge zur Epidemiologie und Physiologie des parasitischen Copepoden *Mytilicola intestinalis*." *Arch. Fischereiwiss.* 2: 120–34.

Millar, R.H. 1963. "Oysters killed by trematode parasites." *Nature* 197: 616.

Miller, R.S. 1967. "Pattern and process in competition." *Adv. Ecol. Res.* 4: 1–74.

Mittal, A.K., Whitear, M. and Agarwal, S.G. 1980. "Fine structure and histochemistry of the epidermis of the fish *Monopterus cuchia*." *J. Zool.* 191: 107–125.

Mizelle, J.D., LaGrave, D.R. and O'Shaughnessy, R.P. 1943. "Studies on monogenetic trematodes. IX. Host specificity of *Pomoxis* Tetraonchinae." *Am. Midland Naturalist* 29: 730–31.

Möller, H. 1976. "Reduction of the intestinal parasite fauna of marine fishes in captivity." *J. mar. biol. Ass. U.K.* 56: 781–85.
Moore, M.N., Lowe, D.M. and Gee, J.M. 1978. "Histopathological effects induced in *Mytilus edulis* by *Mytilicola intestinalis* and the histochemistry of the copepod intestinal cell." *J. Cons. int. Explor. Mer.* 38: 6–11.
Moriya, K. 1944. "On a new species of Aspidocotylea, *Lophotaspis corbiculae*." *Sc. Rep. Hiroshima Bunrika Daigaku*, sect. zool. 10: 231–48.
Murray, M.D. 1976. "Insect parasites of marine birds and mammals." In: Cheng, L. ed.: *Marine Insects*. Amsterdam, Oxford: North-Holland Publ. C. 79–96.
Myers, B.J. 1979. "Anisakine nematodes in fresh commercial fish from waters along the Washington, Oregon and California coasts." *J. Food Protection* 42: 380–84.
Natarajan, P. and James, P.S.B.R. 1977. "A bibliography of parasites and diseases of marine and freshwater fishes of India." *J. Fish Biol.* 10: 347–69.
Needham, T. and Wootten, R. 1978. "The parasitology of teleosts." In: Roberts, R.J. ed.: *Fish Pathology*. London: Baillière Tindal, 144–82.
Negus, M.R.S. 1968. "The nutrition of sporocysts of the trematode *Cercaria doricha* Rothschild, 1935 in the molluscan host *Turritella communis* Risso." *Parasitol.* 58: 355–66.
Nicoll, W. 1914. "The trematode parasites of fishes from the English Channel." *J. mar. biol. Ass. U.K.* 10: 466–505.
Nigrelli, R.F. 1935a. "Experiments on the control of *Epibdella melleni* MacCallum, a monogenetic trematode of marine fishes." *J. Paras.* 21: 438.
Nigrelli, R.F. 1935b. "On the effect of fish mucus on *Epibdella melleni*, a monogenetic trematode of marine fishes." *J. Paras.* 21: 438.
Nigrelli, R.F. 1937. "Further studies on the susceptibility and acquired immunity of marine fishes to *Epibdella melleni*, a monogenetic trematode." *Zoologica* 22: 185–92.
Nigrelli, R.F. 1947. "Susceptibility and immunity of marine fishes to *Benedenia* (= *Epibdella*) *melleni* (MacCallum), a monogenetic trematode. III. Natural hosts in the West Indies." *J. Paras.* 33. Suppl.
Nigrelli, R.F. and Breder, C.H. jr. 1934. "The susceptibility and immunity of certain marine fishes to *Epibdella melleni*, a monogenetic trematode. *J. Paras.* 20: 259–69.
Noble, E.R. 1966. "Myxosporida in deepwater fishes." *J. Paras.* 52: 685–90.
Noble, E.R. 1973. "Parasites and fishes in a deep-sea environment." *Adv. mar. Biol.* 11: 121–95.
Noble, E.R. and Collard, S.B. 1970. "The parasites of midwater fishes. In: Snieszko, S.F. ed.: *A Symposium on Diseases of Fishes and Shellfishes*. Amer. Fish. Soc., Special Publ. 5, Washington, D.C. 57–68.
Noble, E.R., King, R.E. and Jacobs, B.L. 1963. "Ecology of gill parasites of *Gillichthys mirabilis* Cooper." *Ecology* 44: 295–305.
Noble, E.R. and Noble, G.A. 1976. *Parasitology. The Biology of Animal Parasites*. Philadelphia: Lea and Febiger.

Nollen, P.M. 1968. "Autoradiographic studies on reproduction in *Philophthalmus megalurus* (Cort, 1914) (Trematoda). *J. Paras.* 54: 43–48.
Odening, K. 1969. *Entwicklungswege der Schmarotzerwürmer oder Helminthen.* Leipzig: Geest and Portig.
Odening, K. 1974. *Parasitismus, Grundfragen und Grundbegriffe.* Braunschweig: Vieweg und Sohn.
Odening, K. 1976. "Conception and terminology of hosts in parasitology." *Adv. Paras.* 14: 1–93.
Odlaug, T.O. 1946. "The effect of the copepod, *Mytilicola orientalis* upon the Olympia oyster, *Ostrea lurida*." *Trans. Amer. Microsc. Soc.* 65: 311–17.
Olsen, O.W. and Lyons, E.T. 1965. "Life cycle of *Uncinaria lucasi* Stiles 1901 (Nematoda: Ancylostomatidae) of fur seals, *Callorhinus ursinus* Linn., on the Pribilof Islands, Alaska." *J. Paras.* 51: 689–700.
Olsen, R.E. 1978. "Parasitology of the English sole, *Parophrys vetulus* Girard in Oregon, U.S.A." *J. Fish. Biol.* 13: 237–48.
Olsen, R.E. and Pratt, I. 1971. "The life cycle and larval development of *Echinorhynchus lageniformis* Ekbaum, 1938 (Acanthocephala: Echinorhynchidae)." *J. Paras.* 57: 143–49.
Olsen, R.E. and Pratt, I. 1973. "Parasites as indicators of English sole (*Parophrys vetulus*) nursery grounds." *Trans. Amer. Fish. Soc.* 102: 405–11.
O'Meara, G.F. 1976. "Saltmarsh mosquitoes (Diptera: Culicidae)." In: Cheng, L. ed.: *Marine Insects.* Amsterdam, Oxford: North-Holland Publ. Co., 303–33.
Oosten, J. van. 1957. "The skin and scales." In: Brown, M.E. ed.: *The Physiology of Fishes*, vol. 1. New York: Academic Press: 207–44.
Orias, J.D., Noble, E.R. and Alderson, G.D. 1978. "Parasitism in some East Atlantic bathypelagic fishes with a description of *Lecithophyllum irelandeum* sp. n. (Trematoda)." *J. Paras.* 64: 49–51.
O'Rourke, J.F. 1961. "Presence of blood antigens in fish mucus and its possible parasitological significance." *Nature* 189: 943.
Osche, G. 1966. *Die Welt der Parasiten.* Berlin–Heidelberg–New York: Springer.
Osmanov, S.O. 1940. "(Materials on the parasite fauna of fish in the Black Sea.)" *Sc. Mem. Herzen State Pedag. Inst.*, div. zool. 30: 187–265 (in Russian).
Overstreet, R.M. 1976. "*Fabespora vermicola* sp. n., the first myxosporidian from a platyhelminth." *J. Paras.* 62: 680–84.
Overstreet, R.M. and Hochberg, F.G. jr. 1975. "Digenetic trematodes in cephalopods." *J. mar. biol. Ass. U.K.* 55: 893–910.
Owen, H.M. 1957. "Etiological studies on oyster mortality. II. *Polydora websteri* Hartmann (Polychaeta: Spionidae)." *Bull. Mar. Sc. of the Gulf and Caribbean* 7: 35–46.
Paling, J.E. 1965. "The population dynamics of the monogenean gill parasite *Discocotyle sagittata* Leuckart on Windermere trout, *Salmo trutta. L.*" Parasit. 55: 667–94.
Paperna, I. 1963. "Dynamics of *Dactylogyrus vastator* Nybelin (Monogenea)

populations on the gills of carp fry in fish ponds." *Bamidget Bull. Fish Culture*, Israel. 15: 31–50.

Paperna, I. 1975. "Parasites and diseases of the grey mullet (Mugilidae) with special reference to the seas of the Near East." *Aquaculture* 5: 65–80.

Paperna, I. 1978. "Occurrence of fatal parasitic epizootics in maricultured tropical fish." In: *4th Inter. Congr. Paras.* 16–26 August, 1978. Warsaw. Section C., 198.

Paperna, I. and Zwerner, D.E. 1974. "*Kudoa cerebralis* sp. n. (Myxosporidea: Chloromyxidae) from the Striped Bass, *Morone saxatilis* (Walbaum)." *J. Protozool.* 21: 15–19.

Paperna, I. and Zwerner, D.E. 1976a. "Studies on *Ergasilus labracis* Krøyer (Cyclopidea: Ergasilidae) parasitic on striped bass, *Morone saxatilis*, from the lower Chesapeake Bay. I. Distribution, life cycle, and seasonal abundance." *Can. J. Zool.* 54: 449–62.

Paperna, I, and Zwerner, D.E. 1976b. "Parasites and diseases of striped bass, *Morone saxatilis* (Walbaum), from the lower Chesapeake Bay." *J. Fish. Biol.* 9: 267–87.

Parukhin, A.M. 1975 "(On distribution in the world ocean of Nematoda found in fish from the southern seas.)" *Vestnik Zool.* 1, 33–38 (in Russian).

Parukhin, A. and Todorov, I. 1977. "Parasitäre Invasion bei Fischen aus dem Atlantischen Ozean." *Schlachten und Vermarkten* 78: 174.

Pauley, G.B. and Sparks, A.K. 1965. "Preliminary observations on the acute inflammatory reaction in the Pacific Oyster, *Crassostrea gigas* (Thunberg)." *J. Invertebr. Pathol.* 7: 248–56.

Pauley, G.B., Sparks, A.K., Chew, K.K. and Robbins, E.J. 1966. "Infection in Pacific Coast mollusks by thigmotrichid ciliates." *Proc. Nat. Shellfisheries Ass.* 56: 8.

Payne, W.L., Gerding, T.A., Dent, R.G., Bier, J.W. and Jackson, G.J. 1980. "Survey of the U.S. Atlantic Coast surf clam, *Spisula solidissima*, and clam products for anisakine nematodes and hyperparasitic protozoa." *J. Paras.* 66: 150–53.

Pearce, J.B. 1966. "The biology of the mussel crab, *Fabia subquadrata*, from the waters of the San Juan Archipelago, Washington." *Pacific Sci.* 20: 3–35.

Pearse, A.S. and Wharton, G.W. 1938. "The oyster 'leech', *Stylochus inimicus* Palombi, associated with oysters on the coasts of Florida." *Ecol. Monogr.* 8: 605–655.

Pearre, S. jr. 1976. "Gigantism and partial parasitic castration of Chaetognatha infected with larval trematodes." *J. mar. biol. Ass. U.K.*: 56: 503–13.

Pearson, J.C. 1968. "Observations on the morphology and life-cycle of *Paucivitellosus fragilis* Coil, Reid and Kuntz 1965 (Trematoda: Bivesiculidae)." *Parasitol.* 58: 769–88.

Perkins, F.O. 1968. "Fine structure of the oyster pathogen *Minchinia nelsoni* (Haplosporida, Haplosporidiidae)." *J. Invertebr. Pathol.* 10: 287–307.

Perkins, F.O. 1971. "Sporulation in the trematode hyperparasite *Urospo-*

ridium crescens DeTurk, 1940 (Haplosporida: Haplosporidiidae)—an electron microscope study." *J. Paras.* 57: 9–23.

Perkins, F.O. and Wolf, P.H. 1976. "Fine structure of *Marteilia sydneyi* sp. n.—Halplosporidan pathogen of Australian oysters." *J. Paras.* 62: 528–38.

Perkins, F.O., Zwerner, D.E. and Dias, R.K. 1975. "The hyperparasite, *Urosporidium spisuli* sp. n. (Haplosporea), and its effects on the surf clam industry." *J. Paras.* 61: 944–49.

Petrushevski, G.K. and Shulman, S.S. 1961. "The parasitic disease of fishes in the natural waters of the USSR." In: Dogiel, V.A., Petrushevski, G.K. and Polyanski, Yu.I. ed.: *Parasitology of Fishes*. English transition Edinburgh–London: Oliver and Boyd, 299–319.

Pflugfelder, O. 1977. *Wirtstierreaktionen auf Zooparasiten*. Jena: VEB Gustav Fischer.

Phillips, W.J. and Cannon, L.R.G. 1978. "Ecological observations on the commercial sand crab, *Portunus pelagicus* (L.), and its parasite, *Sacculina granifera* Boschma, 1973 (Cirripedia: Rhizocephala)." *J. Fish. Dis.* 1: 137–49.

Pianka, F.R. 1976. "Competition and niche theory." In: May, R.M. ed.: *Theoretical Ecology: Principles and Applications*. Oxford, London, Edinburgh, Melbourne: Blackwell Sc. Publ., 114–41.

Pickering, A.D. 1977. "Seasonal changes in the epidermis of the brown trout *Salmo trutta* (L.)." *J. Fish. Biol.* 10: 561–66.

Pickering, A.D. and Macey, D.J. 1977. "Structure, histochemistry and the effect of handling of the mucous cells of the epidermis of the char *Salvelinus alpinus* (L.)." *J. Fish. Biol.* 10: 505–12.

Platt, P. 1968. "The effect of endoparasitism by *Cryptocotyle lingua* (Creplin) on digestion in the snail, *Littorina littorea*." M. Sc. thesis, Dalhousie University, Halifax, N.S.

Platt, N.E. 1975. "Infestation of cod (*Gadus morhua* L.) with larvae of codworm (*Terranova decipiens* Krabbe) and herringworm, *Anisakis* sp. (Nematoda: Ascaridata), in North Atlantic and Arctic waters." *J. Appl. Ecol.* 12: 437–50.

Platt, N.E. 1976. "Codworm—a possible biological indicator of the degree of mixing of Greenland and Iceland cod stocks." *J. Conseil. Inter. pour l'Exploration de la Mer.* 37: 41–45.

Pohley, W.J. 1976. "Relationship among three species of *Littorina* and their larval Digenea." *Mar. Biol.* 37: 179–86.

Polyanski, Yu.I. 1955. "(Materials on the parasitology of fish of the northern seas of the USSR.)" *Trud. Zool. Inst. Akad. Nauk SSSR* 19: 5–170 (in Russian).

Polyanski, Yu.I. 1961a. "Ecology of parasites of marine fishes." In: Dogiel, V.A., Petrushevski, G.K. and Polyanski, Yu.I. *Parasitology of Fishes*. English translation Edinburgh, London: Oliver and Boyd, 48–83.

Polyanski, Yu.I. 1961b. "Zoogeography of the parasites of USSR marine fishes." In: Dogiel, V.A., Petrushevski, G.K. and Polyanski, Yu.I.: *Parasitology of Fishes*. English translation Edinburgh, London: Oliver and Boyd. 230–45.

Polyanski, Yu.I. 1966. *Parasites of the Fish of the Barents Sea.* English translation Jerusalem: I.P.S.T.

Potts, G.W. 1968. "The ethology of *Crenilabrus melanocercus*, with notes on cleaning symbiosis." *J. mar. biol. Ass. U.K.* 48: 279–93.

Potts, G.W. 1973a. "Cleaning symbiosis among British fish with special reference to *Crenilabrus melops* (Labridae)." *J. mar. biol. Ass. U.K.* 53: 1–10.

Potts, G.W. 1973b. "The ethology of *Labroides dimidiatus* (Cuv. & Val.) (Labridae, Pisces) on Aldabra." *Animal Behaviour* 21: 250–91.

Pratt, I. and Herrman, R. 1962. "*Nitzschia quadritestes* sp. n. (Monogenea: Capsalidae) from the Columbia river sturgeon." *J. Paras.* 48: 291–92.

Price, P.W. 1977. "General concepts on the evolutionary biology of parasites." *Evolution* 31: 405–20.

Price, P.W. 1980. *Evolutionary Biology of Parasites.* Princeton, New Jersey: Princeton University Press.

Prytherch, H.F. 1940. "The life cycle and morphology of *Nematopsis ostrearum*, sp. nov. a gregarine parasite of the mud crab and oyster." *J. Morphol.* 66: 39–65.

Putz, R.E. and Hoffman, G.L. 1964. "Studies on *Dactylogyrus corporalis* n. sp. (Trematoda Monogenea) from the fallfish *Semotilus corporalis*." *Proc. Helm. Soc. Wash.* 31: 139–43.

Quayle, D.B. 1975. *Tropical Oyster Culture. A Selected Bibliography.* Ottawa, Canada; Inter. Development Res. Centre. 40 pp.

Rae, B.B. 1972. *A review of the cod-worm problem in the North Sea and in western Scottish waters 1958–1970.* Marine Res. Dept. Agr. Fisheries, Scotland, No. 2. 24 pp.

Rajbanshi, V.K. 1977. "The architecture of the gill surface of the catfish, *Heteropneustes fossilis* (Bloch): SEM study." *J. Fish. Biol.* 10: 325–29.

Ramalingam, K. 1960. "The functional development of 'compensating asymmetry' in the higher Monogenea." *Z. wiss Zool.* 164: 347–81.

Rand, A.S. 1969. "Remarks on aquatic habits of the Galapagos marine iguana, including submergence times, cleaning symbiosis, and the shark threat." *Copeia*, 1969. 2: 401–402.

Randall, J.E. 1958. "A review of the labrid fish genus *Labroides*, with descriptions of two new species and notes on ecology." *Pacif. Sc.* 12: 327–47.

Randall, J.E. 1962. "Fish service stations." *Sea Frontiers* 8: 40–47.

Rawson, M.V. jr. 1973. "The development and seasonal abundance of the parasites of striped mullet, *Mugil cephalus* L., and mummichogs, *Fundulus heteroclitus* (L.)." Ph.D. thesis, University of Georgia, Athens, Georgia. 99 pp.

Read, C.P. 1958. "A science of symbiosis." *Amer. Inst. Biol. Sci. Bull.* 8: 16–17.

Rees, B. 1960. "*Albertia vermicularis* (Rotifera) parasitic in the earthworm *Allolobophora caliginosa*." *Parasitol.* 50: 61–66.

Rees, G. 1951. "The anatomy of *Cysticercus taeniae-taeniaeformis* (Batsch, 1786) (*Cysticercus fasciolaris* Rud. 1808), from the liver of *Rattus norvegicus* (Erx.), including an account of spiral torsion in the species and some minor abnormalities in structure." *Parasitol.* 41: 46–59.

Rees, W.J. 1967. "A brief survey of the symbiotic associations of the Cnidaria with Mollusca." *Proc. malac. Soc. Lond.* 37: 213–31.

Reichenbach-Klinke, H.H. 1973. "Der Einfluss des Parasitenbefalls auf Abwachs und Eizahl der Fische." *Verh. Internat. Verein. Limnol.* 18: 1639–48.

Reichenbach-Klinke, H.H. 1975. "Die Bedeutung der Parasiten für die Produktion von Süsswasserfischen." *Fisch und Umwelt* 1: 113–21:

Reichenbach-Klinke, H.H. and Elkan, H. 1965. *The Principal Diseases of Lower Vertebrates*. London: Academic Press.

Reichenbach-Klinke, H.H. and Landolt, N. 1973. *Fish Pathology*. Hong Kong: T.F.H. Publ.

Reinhard, E.G. 1956. "Parasitic castration of Crustacea." *Exp. Parasitol.* 5: 79–107.

Remley, L.W. 1936. "Morphology and life-history studies of *Microcotyle spinicirrus*, a monogenetic trematode from the gills of *Aplodinotus grunniens*." *J. Paras.* 22: suppl. 535.

Remley, L.W. 1942. "Morphology and life history studies of *Microcotyle spinicirrus* MacCallum, 1918, a monogenetic trematode parasitic on the gills of *Aplodinotus grunniens*." *Trans. Am. micr. Soc.* 61: 141–55.

Reshetnikova, A.V. 1955. "(Contributions to parasite fauna of the fishes of the Black Sea.)" *Tr. Karadagsk. biol. st.* 13 (in Russian).

Ridgway, S.H., Geraci, J.R. and Medway, W. 1975. "Diseases of pinnipeds." *Internat. J. Expl. of the Sea* 169: 327–37.

Rifkin, E. and Cheng, T.C. 1968. "The origin, structure, and histochemical characterization of encapsulating cysts in the oyster *Crassostrea virginica* parasitized by the cestode *Tylocephalum* sp." *J. Invertebr. Pathol.* 10: 54–64.

Roberts, R.J., Young, H. and Milne, J.A. 1971. "Studies on the skin of plaice (*Pleuronectes platessa* L.) I. The structure and ultrastructure of normal plaice skin." *J. Fish. Biol.* 4: 87–98.

Robertson, R. 1973. "Sex changes under the sea." *New Scientist* 58: 538–40.

Robinson, E.S. 1961. "Some monogenetic trematodes from marine fishes of the Pacific." *Trans. Am. Microsc. Soc.* 80: 235–66.

Rogers, W.P. 1957. "An alternative approach to the study of host–parasite specificity." *Intern. Union Biol. Sc. Coll.* 32: 309–11.

Rogers, W.P. 1962. *The Nature of Parasitism*. New York and London: Acad. Press.

Rogers, W.P. 1963. "Physiology of infection with nematodes: some effects of the host stimulus on infective stages." *Ann. N.Y. Acad. Sc.* 113: 208–16.

Rohde, K. 1966. "Sense receptors of *Multicotyle purvisi* Dawes, (Trematoda, Aspidobothria)." *Nature* 211: 820–22.

Rohde, K. 1968a. "The nervous systems of *Multicotyle purvisi* Dawes, 1941. (Aspidogastrea) and *Diaschistorchis multitesticularis* Rohde, 1962 (Digenea)." *Z. Parasitenk.* 30: 78–94.

Rohde, K. 1968b. "Das Nervensystem der Gattung *Polystomoides* Ward, 1917 (Monogenea)." *Z. Morph. Tiere* 62: 58–76.

Rohde, K. 1968c. "Das Nervensystem der Cercarie von *Catatropis indica* Srivastava, 1935 (Digenea: Notocotylidae) und der geschlechtsreifen Form von *Diaschistorchis multitesticularis* Rohde, 1962 (Digenea: Pronocephalidae)." *Z. Morph. Tiere* 62: 77–102.
Rohde, K. 1968d. "Lichtmikroskopische Untersuchungen an den Sinnesrezeptoren der Trematoden." *Z. Parasitenk.* 30: 252–77.
Rohde, K. 1970. "Nerve sheath in *Multicotyle purvisi* Dawes." *Naturwissenschaften* 57: 502–03.
Rohde, K. 1971. "Untersuchungen an *Multicotyle purvisi* Dawes, 1941 (Trematoda: Aspidogastrea). III. Licht- und elektronenmikroskopischer Bau des Nervensystems." *Zool. Jb. Anat.* 88: 320–63.
Rohde, K. 1972a. "The Aspidogastrea, especially *Multicotyle purvisi* Dawes, 1941." *Adv. Paras.* 10: 77–151.
Rohde, K. 1972b. "Sinnesrezeptoren von *Lobatostoma* n. sp. (Trematoda: Aspidogastrea)." *Naturwissenschaften* 59: 168.
Rohde, K. 1972c. "Ultrastructure of the nerves and sense receptors of *Polystomoides renschi* Rohde and *P. malayi* Rohde (Monogenea: Polystomatidae)." *Z. Parasitenk.* 40: 307–20.
Rohde, K. 1972d. "Die Entwicklung von *Lobatostoma* n. sp. (Trematoda, Aspidogastrea)." *Naturwissenschaften* 59: 168.
Rohde, K. 1973. "Structure and development of *Lobatostoma manteri* sp. nov. (Trematoda: Aspidogastrea) from the Great Barrier Reef, Australia." *Parasitol.* 66: 63–83.
Rohde, K. 1975a. "Fine structure of the Monogenea, especially *Polystomoides* Ward." *Adv. Parasit.* 13: 1–33.
Rohde, K. 1975b. "Early development and pathogenesis of *Lobatostoma manteri* Rohde (Trematoda: Aspidogastrea)." *Int. J. Parasit.* 5: 597–607.
Rohde, K. 1976a. "Species diversity of parasites on the Great Barrier Reef." *Z. Parasitenk.* 50: 93–94.
Rohde, K. 1976b. "Monogenean gill parasites of *Scomberomorus commersoni* Lacépède and other mackerel on the Australian east coast." *Z. Parasitenk.* 51: 49–69.
Rohde, K. 1976c. "Marine parasitology in Australia." *Search* 7: 477–82.
Rohde, K. 1977a. "Habitat partitioning in Monogenea of marine fishes. *Heteromicrocolyla australiensis*, sp. nov. and *Heteromicrocotyloides mirabilis*, gen. and sp. nov. (Heteromicrocotylidae) on the gills of *Carangoides emburyi* (Carangidae) on the Great Barrier Reef, Australia." *Z. Parasitenk.* 53: 171–82.
Rohde, K. 1977b. "Species diversity of monogenean gill parasites of fish on the Great Barrier Reef." *Proc. Third Internat. Coral Reef Symposium—Miami, Florida*, 585–91.
Rohde, K. 1977c. "A non-competitive mechanism responsible for restricting niches." *Zool. Anz.* 199: 164–72.
Rohde, K. 1977d. "The bird schistosome *Austrobilharzia terrigalensis* from the Great Barrier Reef, Australia." *Z. Parasitenk.* 52: 39–51.
Rohde, K. 1978a. "Monogenea of Australian marine fishes. The genera

Dionchus, Sibitrema and *Hexostoma*." *Publ. Seto. Mar. Biol. Lab.* 24: 349–67.

Rohde, K. 1978b. "Monogenean gill parasites of the kingfish *Seriola grandis* (Castlenau) from the Great Barrier Reef." *Publ. Seto. Mar. Biol. Lab.* 24: 369–76.

Rohde, K. 1978c. "The bird schistosome *Gigantobilharzia* sp. in the silver gull, *Larus novaehollandiae*, a potential agent of schistosome dermatitis in Australia." *Search* 9: 40–42.

Rohde, K. 1978d. "Latitudinal gradients in species diversity and their causes. I. A review of the hypotheses explaining the gradients." *Biol. Zentralbl.* 97: 393–403.

Rohde, K. 1978e. "Latitudinal gradients in species diversity and their causes. II. Marine parasitological evidence for a time hypothesis." *Biol. Zentralbl.* 97: 405–18.

Rohde, K. 1978f. "Latitudinal differences in host specificity of marine Monogenea and Digenea." *Mar. Biol.* 47: 125–34.

Rohde, K. 1979a. "The buccal organ of some Monogenea Polyopisthocotylea." *Zoologica Scripta* 8: 161–70.

Rohde, K. 1979b. "Monogenean gill parasites of some marine fishes of Papua New Guinea." *Zool. Anz.* 203: 78–94.

Rohde, K. 1979c. "A critical evaluation of intrinsic and extrinsic factors responsible for niche restriction in parasites." *Amer. Natural.* 114: 648–71.

Rohde, K. 1980a. "Warum sind ökologische Nischen begrenzt? Zwischenartlicher Antagonismus oder innerartlicher Zusammenhalt?" *Naturwissenschaftliche Rundschau* 33: 98–102.

Rohde, K. 1980b. "Host specificity indices of parasites." *Proc. 24th Conf. Austr. Soc. Paras.* Adelaide, May.

Rohde, K. 1980c. "Some aspects of the ultrastructure of *Gotocotyla secunda* (Tripathi, 1954) (Monogenea: Gotocotylidae) and *Hexostoma euthynni* Merserve, 1938 (Monogenea: Hexostomatidae)." *Angewandte Parasitologie* 21: 32–48.

Rohde, K. 1980d. "Comparative studies on microhabitat utilization by ectoparasites of some marine fishes from the North Sea and Papua New Guinea." *Zool. Anz.* 204: 27–64.

Rohde, K. 1980e. "Host specificity indices of parasites and their application." *Experientia* 36: 1369–71.

Rohde, K. 1980f. "Diversity gradients of marine Monogenea in the Atlantic and Pacific Oceans." *Experientia* 36: 1368–69.

Rohde, K. 1981a. "Niche width of parasites in species-rich and species-poor communities." *Experientia* 37: 359–61.

Rohde, K. 1981b. "Population dynamics of two snail species, *Planaxis sulcatus* and *Cerithium moniliferum*, and their trematode species at Heron Island, Great Barrier Reef." *Oecologia* 49: 344–352.

Rohde, K., Roubal, F. and Hewitt, G.C. 1980. "Ectoparasitic Monogenea, Digenea and Copepoda from the gills of some marine fishes of New Caledonia and New Zealand." *New Zealand J. Mar. Freshwater Res.* 14: 1–13.

Rohde, K. and Sandland, R. 1973. "Host–parasite relations in *Lobatostoma manteri* Rohde (Trematoda: Aspidogastrea)." *Z. Parasitenk.* 42: 115–36.

Rothschild, M. 1936. "Gigantism and variation in *Peringia ulvae* Pennant 1777, caused by infection with larval trematodes." *J. mar. biol. Ass. U.K.* 20: 537–46.

Rothschild, M. 1938. "Further observations on the effect of trematode parasites on *Peringia ulvae* (Pennant) 1777." *Novitates Zoologicae* 41: 84–102.

Rothschild, M. 1941a. "The effect of trematode parasites on the growth of *Littorina neritoides* (L.)." *J. mar. biol. Ass. U.K.* 25: 69–80.

Rothschild, M. 1941b. "Observations on the growth and trematode infections of *Peringia ulvae* (Pennant 1777) in a pool in the Tamar saltings, Plymouth." *Parasitol.* 33: 406–415.

Roubal, F. 1979. "Taxonomy and site specificity of the ectoparasitic metazoans on the Black Bream, *Acanthopagrus australis* in northern New South Wales." M. Sc. thesis, University of New England, Armidale.

Roughley, T.C. 1922. "Oyster culture on the George's River, New South Wales." *Technol. Educ. Ser.* 25, Technol. Mus., Sydney. 1–69.

Roughley, T.C. 1925. "The story of the oyster." *Aust. Mus. Mag.* 2: 1–32.

Roughley, T.C. 1926. "An investigation of the cause of an oyster mortality on the Georges River, New South Wales, 1924–5." *Proc. Linn. Soc. N.S.W.* 51: 446–90.

Roughley, T.C. 1951. *Fish and Fisheries of Australia.* Sydney: Angus and Robertson Ltd.

Rützler, K. and Rieger, G. 1973. "Sponge burrowing: fine structure of *Cliona lampa* penetrating calcareous substrate." *Mar. Biol.* 21: 144–62.

Sale, P.F., McWilliam, P.S. and Anderson, D.T. 1976. "Composition of the near-reef zooplankton at Heron Reef, Great Barrier Reef." *Mar. Biol.* 34: 59–66.

Sandars, D.F. 1957. "Redescription of some cestodes from marsupials. I. Taeniidae." *Ann. Trop. Med. Parasit.* 51: 317–29.

Sanders, M.J. 1966. "Parasitic castration of the scallop *Pecten alba* (Tate) by a bucephalid trematode." *Nature* 212: 307–308.

Sankurathri, C.S. and Holmes, J.C. 1976. "Effects of thermal effluents on parasites and commensals of *Physa gyrina* Say (Mollusca: Gastropoda) and their interactions at Lake Wabamun, Alberta." *Can. J. Zool.* 54: 1742–53.

Sastry, A.N. and Menzel, R.W. 1962. "Influence of hosts on the behaviour of the commensal crab *Pinnotheres maculatus* Say." *Biol. Bull.* 13: 388–95.

Schäperclaus, W. 1954. *Fischkrankheiten.* Berlin: Akademie Verlag.

Schell, S.C. 1972. "The early development of *Udonella caligorum* Johnston, 1835 (Trematoda: Monogenea)." *J. Paras.* 58: 1119–21.

Schell, S.C. 1973. "*Rugogaster hydrolagi* gen. et sp. n. (Trematoda: Aspidobothrea: Rugogastridae fam. n.). from the ratfish, *Hydrolagus colliei* (Lay and Bennett 1839)." *J. Paras.* 59: 803–805.

Scheltema, R.S. 1962. "The relationship between the flagellate protozoon *Hexamita* and the oyster *Crassostrea virginica*." *J. Paras.* 48: 137–41.

Schlicht, F.G. and Mackin, J.G. 1968. "*Hexamita nelsoni* sp. n. (Polymastigina: Hexamitidae) parasitic in oysters." *J. Invertebr. Pathol.* 11: 35–39.

Schuhmacher, H. 1973. "Das kommensalische Verhältnis zwischen *Periclimenes imperator* (Decapoda: Palaemonidae) und *Hexabranchus sanguineus* (Nudibranchia: Doridacea)." *Mar. Biol.* 22: 355–60.

Shipley, A.E. and Hornell, J. 1904. "The parasites of the pearl oyster." *Rep. Gov. Ceylon on Pearl Oyster Fisheries Gulf of Manaar.* pt. 2: 77–106.

Shiraki, T. 1974. "Larval nematodes of family Anisakidae (Nematoda) in the northern Sea of Japan—as a causative agent of eosinophilic phlegmone or granuloma in the human gastro-intestinal tract." *Acta Medica et Biologica* 22: 57–98.

Short, R.B. 1962. "Two new dicyemid mesozoans from the Gulf of Mexico." *Tulane Studies in Zool.* 9: 101–111.

Shotter, R.A. 1976. "The distribution of some helminth and copepod parasites in tissues of whiting, *Merlangius merlangus* L., from Manx waters." *J. Fish. Biol.* 8: 101–117.

Shulman, S.S. and Shulman-Albova. R.E. 1953. (*Parasites of Fishes of the White Sea.*) Moscow, Leningrad: Isd. Akad. Nauk SSSR (in Russian).

Shuster, C.N. and Hillman, R.E. 1963. "Comments on 'Microecological Factors on Oyster Epizootics' by Marshall Laird." *Chesapeake Sc.* 4: 101–103.

Simmons, J.E. and Laurie, J.S. 1972. "A study of *Gyrocotyle* in the San Juan Archipelago, Puget Sound, U.S.A., with observations on the host, *Hydrolagus colliei* (Lay and Bennett)." *Int. J. Parasit.* 2: 59–77.

Sindermann, C.J. 1957. "Diseases of fishes of the western north Atlantic V. Parasites as indicators of herring movements." *Maine Dept. Sea and Shore Fisheries, Res. Bull.* No. 27: 1–30.

Sindermann, C.J. 1963. "Disease in marine populations." *Trans. North Amer. Wildlife Nat. Res. Conf.* 28: 336–56.

Sindermann, C.J. 1966. "Diseases of marine fishes." *Adv. Mar. Biol.* 4: 1–89.

Sindermann, C.J. 1970. *Principal Diseases of Marine Fish and Shellfish.* New York and London: Acad. Press.

Sindermann, C.J. (Editor). 1977. *Disease Diagnosis and Control in North American Marine Aquaculture. Developments in Aquaculture and Fisheries Science.* No. 6. Amsterdam: Elsevier Sc. Publ. Co.

Sindermann, C.J. 1978. ed.: *Disease Diagnosis and Control in North American Marine Aquaculture. Developments in Aquaculture and Fisheries Science.* No. 6. Amsterdam: Elsevier Sc. Publ. Co.

Sindermann, C.J. and Rosenfield, A. 1967. "Principal diseases of commercially important marine bivalve Mollusca and Crustacea." *U.S. Fish Wildlife Serv., Fishery Bull.* 66: 335–85.

Slobodkin, L.B. and Fishelson, L. 1974. "The effect of the cleaner fish *Labroides dimidiatus* on the point diversity of fishes in the reef found at Eilat." *Amer. Natural.* 108: 369–76.

Sluiters, J.F. 1974. "*Anisakis* sp. larvae in the stomachs of herrings (*Clupea harengus* L.)." *Z. Parasitenk.* 44: 279–88.
Smith, G.F.M. 1936. "A gonad parasite of the starfish." *Science* 148: 157.
Smith, J.W. and Wootten, R. 1975. "Experimental studies on the migration of *Anisakis* sp. larvae (Nematoda: Ascaridida) into the flesh of herring, *Clupea harengus* L." *Int. J. Paras.* 5: 133–36.
Smith, J.W. and Wootten, R. 1978. "*Anisakis* and anisakiasis." *Adv. Paras.* 16: 93–163.
Smyth, J.D. 1962. *Introduction to Animal Parasitology.* Springfield, Ill.: Charles C. Thomas.
Snieszko, S.F. ed. 1970. *A Symposium on Diseases of Fishes and Shellfishes.* Amer. Fish. Soc., Special Publ. 5, Washington, D.C.
Sogandares-Bernal, F. 1959. "Digenetic trematodes of marine fishes from the Gulf of Panama and Bimini, British West Indies." *Tulane Stud. Zool.* 7: 70–117.
Sommerville, R.I. 1957. "The exsheathing mechanism of nematode infective larvae." *Exp. Parasit.* 6: 18–30.
Southwell, T. 1924. "The pearl-inducing worm in the Ceylon pearl oyster." *Ann. Trop. Med. Paras.* 18: 37–53.
Sparks, A.K. 1962. "Metaplasia of the gut of the oyster *Crassostrea gigas* (Thunberg) caused by infection with the copepod *Mytilicola orientalis* Mori." *J. Insect Pathol.* 4: 57–62.
Sparks, A.K. and Chew, K.K. 1966. "Gross infestation of the Littleneck Clam, *Venerupis staminea*, with a larval cestode (*Echeneibothrium* sp.)." *J. Invertebr. Pathol.* 8: 413–16.
Sprague, V. 1964. "*Nosema dollfusi* n. sp. (Microsporidia; Nosematidae), a hyperparasite of *Bucephalus cuculus* in *Crassostrea virginica*." *J. Protozool.* 11: 381–85.
Sprague, V. 1970. "Some protozoan parasites and hyperparasites in marine bivalve molluscs." In: Snieszko, S.F. ed.: *A Symposium on Diseases of Fishes and Shellfishes.* Amer. Fish. Soc., Special Publ. 5, Washington, D.C.
Sprague, V. and Orr, P.E. jr. 1955. "*Nematopsis ostrearum* and *N. prytherchi* (Eugregarinina: Porosporidae) with special reference to the host–parasite relations." *J. Paras.* 41: 89–104.
Sprent, J.F.A. 1977. "Ascaridoid nematodes of amphibians and reptiles: *Sulcascaris.*" *J. Helminth.* 54: 379–87.
Sproston, N.G. 1946. "A synopsis of monogenetic trematodes." *Tr. Zool. Soc. London.* 25: 185–600.
Stalker, H.D. 1956. "On the evolution of parthenogenesis in Lonchoptera (Diptera)." *Evolution* 10: 345–59.
Stauber, L.A. 1944. "*Pinnotheres ostreum*, parasitic on the American oyster, *Ostrea (Gryphaea) virginica.*" *Biol. Bull.* 87: 269–91.
Stern, L., Myers, B.J., Amish, R.A., Knollenberg, W. and Chew, K. 1976." Anisakine nematodes in commercial marine fish from Washington State." *Trans. Amer. Micro. Soc.* 95: 264.
Strasburg, D.W. 1959. "Notes on the diet and correlating structures of some central pacific echeneid fishes." *Copeiay* 1959, 3: 244–48.

Strasburg, D.W. 1962. "Some aspects of the feeding behaviour of *Remora remora*." *Pacific Sc.* 16: 202–206.
Strasburg, D.W. 1964. "A possible breakdown in symbiosis between fishes." *Copeia*, 1964. 1: 228–29.
Strelkov, Yu.A. 1960. "(Endoparasitic worms of marine fish of eastern Kamchatka.)" *Trud. Zool. Inst. Acad. Nauk SSSR.* 28: 147–96 (in Russian).
Stroud, R.K. and Dailey, M.D. 1978. "Parasites and associated pathology observed in pinnipeds stranded along the Oregon Coast." *J. Wildlife Diseases* 14: 292–98.
Stroud, R.K. and Roffe, T.J. 1979. "Causes of death in marine mammals stranded along the Oregon Coast." *J. Wildlife Diseases* 15: 91–97.
Stunkard, H.W. 1922. "*Caballerocotyla klawei* sp. n., a monogenetic trematode from the nasal capsule of *Neothynnus macropterus*." *J. Paras.* 48: 883–90.
Stunkard, H.W. and Lux, F.E. 1965. "A microsporidian infection of the digestive tract of the winter flounder, *Pseudopleuronectes americanus*." *Biol. Bull.* 129: 371–87.
Stunkard, H.W. and Uzmann, J.R. 1958. "Studies on digenetic trematodes of the genera *Gymnophallus* and *Parvatrema*." *Biol. Bull.* 115: 276–302.
Suomalainen, E. 1962. "Significance of parthenogenesis in the evolution of insects." *Ann. Rev. Entomol.* 7: 349–66.
Sweeney, J.C. 1974. "Common diseases of pinnipeds." *J. Amer. Vet. Med. Assoc.* 165: 805–810.
Sweeney, J.C. and Ridgway, S.H. 1975. "Common diseases of small cetaceans." *J. Amer. Vet. Med. Assoc.* 167: 533–40.
Szidat, L. 1961. "Versuch einer Zoogeographie des Süd-Atlantik mit Hilfe von Leitparasiten der Meeresfische." *Parasit. Schriftenreihe* 13: 1–98.
Szidat, L. and Nani, A. 1951. "Las remoras del Atlantico Austral con un estudio de su nutricion natural y de sus parasitos (Pisc. Echeneidae)." *Rev. Inst. Nacl. Invest. Museo Arg. Cienc. Nat. Zool.* 2: 385–417.
Takatsuki, S. 1934. "On the nature and functions of the amoebocytes of *Ostrea edulis*." *Q. J. microsc. Sci.* 76: 379–431.
Tauber, C.A. and Tauber, M.J. 1977a. "Sympatric speciation based on allelic change at three loci: evidence from natural populations in two habitats." *Science* 197: 1298–99.
Tauber, C.A. and Tauber, M.J. 1977b. "A genetic model for sympatric speciation through habitat diversification and seasonal isolation." *Nature* 268: 702–705.
Tauber, C.A., Tauber, M.J. and Nechols, J.R. 1977. "Two genes control seasonal isolation in sibling species." *Science* 197: 592–93.
Taylor, R.L. 1966. "*Haplosporidium tumefacientis* sp. n., the etiologic agent of a disease of the Californian sea mussel, *Mytilus californianus* Conrad." *J. Invertebr. Pathol.* 8: 109–121.
Thane-Fenchel, A. 1966. "*Proales paguri* sp. nov., a rotifer living on the gills of the hermit crab *Pagurus bernhardus* (L.)." *Ophelia* 3: 93–97.
Thing, H., Clausen, B. and Henriksen, S.A. 1976. "Finding of *Trichinella*

spiralis in a walrus (*Odobenus rosmarus* L.) in the Thule district, northwest Greenland." *Nordisk Veterinaer Medicin* 28: 59.
Thorson, R.E. 1969. "Environmental stimuli and responses of parasitic helminths." *Bioscience* 19: 126–30.
Thresher, R.E. 1977. "Pseudo-cleaning behaviour of Florida reef fishes." *Copeia* 1977: 768–69.
Todorov, I. 1973. "(Occurrence of nematode larvae in some Atlantic fish.)" *Veterinarna Sbirka* 71: 30–32 (in Russian).
Toman, G. 1973. "(Digenetic trematodes of deep sea fish from Suruga Bay (primary report).) *Jap. J. Paras.* 22: 11–12 (in Japanese).
Tomlinson, J. 1966. "The advantages of hermaphroditism and parthenogenesis." *J. Theoret. Biol.* 11: 54–58.
Tripp, M.R. 1958. "Studies on the defense mechanism of the oyster, *Crassostrea virginica.*" *J. Paras.* 44: 35.
Tripp, M.R. 1963. "Cellular responses of mollusks." *Ann. N.Y. Acad. Sc.* 113: 467–74.
Tripp, M.R. 1966. "Hemagglutinin in the blood of the oyster *Crassostrea virginica.*" *J. Invertebrate Pathol.* 8: 478–84.
Tripp, M.R. and Turner, R.M. 1978. "Effects of the trematoda *Proctoeces maculatus* on the mussel *Mytilus edulis.*" In: Bulla, L.A. jr. and Cheng, T.C. eds.: *Comparative Pathobiology.* vol. 4. *Invertebrate Models for Biomedical Research.* New York: Plenum Press.
Turner, C.H., Ebert, E.E. and Given, R.R. 1969. "Man-made reef ecology." *Calif. Dept. Fish Game, Fish. Bull.* 146: 4–221.
Ulmer, M.J. 1970. "Site-finding behaviour in helminths in intermediate and definitive hosts." In: Fallis, A.M. ed.: *Ecology and Physiology of Parasites.* London: Adam Hilger, 123–60.
Uspenskaya, A.V. 1962. "(On the nutrition of monogenetic trematodes.)" *Dokl. Akad. Nauk SSSR* 142: 1212–15 (in Russian).
Vauk, G. 1973. "Beobachtungen am Seehund (*Phoca vitulina* L.) auf Helgoland." *Zeitschrift für Jagdwissenschaft* 19: 117–21.
Vernberg, W.B. and Vernberg, F.J. 1967. "Interrelationships between parasites and their hosts. III. Effect of larval trematodes on the thermal metabolic response of their molluscan host." *Exp. Parasit.* 20: 225–31.
Vevers, H.G. 1951. "The biology of *Asterilas rubens* L. II. Parasitization of the gonads by the ciliate *Orchitophrya stellarum* Cépède." *J. mar. biol. Ass. U.K.* 29: 619–24.
Wakelin, D. 1976. "Host responses." In: Kennedy, C.R. ed.: *Ecological Aspects of Parasitology.* Amsterdam, Oxford: North-Holland Publ. Co., 116–41.
Wahlert, G. von and Wahlert, H. von, 1962. "Beobachtungen und Bemerkungen zum Putzverhalten von Mittelmeerfischen." *Veröff. Bremerhaven Inst. Meeresforschg.* 8: 71–77.
Walters, V. 1955. "Snake-eels as pseudoparasites of fishes." *Copeia,* 1955: 146–47.
Watts, S.D.M. 1971. "Effects of larval Digenea on the free amino acid pool of *Littorina littorea* (L.)." *Parasitol.* 62: 361–66.

Way, M.J. and Cammell, M.E. 1971. "Self regulation in aphid populations." In: den Boer, P.J. and Gradwell, G.R. eds.: *Dynamics of Populations*. *Proc. Adv. Study Inst. on Dynamics of Numbers in Populations*, Oosterbeck, 1970. Centre for Agricultural Publ. Docum. Wageningen, 232–42.

White, I.C. 1972. "On the ecology of an adult digenetic trematode *Proctoeces subtenuis* from a lamellibranch host *Scrobicularia plana*." *J. mar. biol. Ass. U.K.* 52: 457–67.

Whitelegge, T. 1890. "Report on the worm disease affecting the oysters on the coast of New South Wales." *Rec. Aust. Mus.* 1: 41–54.

Whitfield, P.J. 1979. *The Biology of Parasitism* London: Edward Arnold (Publ.) Ltd.

Whitfield, P.J., Anderson, R.M. and Moloney, N.A. 1975. "The attachment of cercariae of an ectoparasitic digenean, *Transversotrema patialensis*, to the fish host: behavioural and ultrastructural aspects." *Parasitol.* 70: 311–29.

Wickler, W. 1960. "Aquarienbeobachtungen an *Aspidontus*, einem ektoparasitischen Fisch." *Z. Tierpsych.* 17: 277–92.

Wickler, W. 1961. "Über das Verhalten der Blenniiden *Runula* und *Aspidontus* (Pisces, Blenniidae)." *Z. Tierpsych.* 18: 421–40.

Wickler, W. 1968. *Mimicry*. New York: World Univ. Library.

Wiles, M. 1968. "The occurrence of *Diplozoon paradoxum* Nordmann, 1832, (Trematoda: Monogenea) in certain waters of northern England and its distribution on the gills of certain Cyprinidae." *Parasitol.* 58: 61–70.

Willey, C.H. and Gross, P.R. 1957. "Pigmentation in the foot of *Littorina littorea* as a means of recognition of infection with trematode larvae." *J. Paras.* 43: 324–27.

Williams, E.H. jr. and Phelps, R.P. 1976. "Parasites of some mariculture fishes before and after cage culture." In: Webber, H.H. and Ruggieri, G.D. eds.: *Food-drugs from the Sea*. Proc. 4th Conf. Washington, U.S.A., Marine Technology Society, 216–30.

Williams, H.H. 1960. "The intestine in members of the genus *Raja* and host-specificity in the Tetraphyllidea." *Nature* 188: 514–16.

Williams, H.H. 1961. "Observations on *Echeneibothrium maculatum* (Cestoda: Tetraphyllidea)." *J. mar. biol. Ass. U.K.* 41: 631–52.

Williams, H.H. 1964. "Some new and little known cestodes from Australian elasmobranchs with a brief discussion on their possible use in problems of host taxonomy." *Parasitol.* 54: 737–48.

Williams, H.H. 1965. "Observations on the occurrence of *Dictyocotyle coeliaca* and *Calicotyle kroyeri* (Trematoda: Monogenea)." *Parasitol.* 55: 201–207.

Williams, H.H. 1966. "The ecology, functional morphology and taxonomy of *Echeneibothrium* Beneden, 1849 (Cestoda: Tetraphyllidea), a revision of the genus and comments on *Discobothrium* Beneden, 1870, *Pseudanthobothrium* Baer, 1956, and *Phormobothrium* Alexander, 1963." *Parasitol.* 56: 227–85.

Williams, H.H. 1967. "Helminth diseases of fish." *Helminth. Abstr.* 36: 261–95.

Williams, H.H. 1968a "The taxonomy, ecology and host-specificity of some Phyllobothriidae (Cestoda: Tetraphyllidea), a critical revision of *Phyllobothrium* Beneden, 1849 and comments on some allied genera." *Phil. Transact. Roy. Soc. London ser. B.* 253: 231–307.
Williams, H.H. 1968b. "*Phyllobothrium piriei* sp. nov. (Cestoda: Tetraphyllidea) from *Raja naevus* with a comment on its habitat and mode of attachment." *Parasitol.* 58: 929–37.
Williams, H.H. and Jones, A. 1976. "Marine helminths and human health." Farnham Royal, Slough, U.K. Commonwealth Agricultural Bureaux. *CIH Miscellaneous Publications* No. 3.
Williams, H.H., McVicar, A.H. and Ralph, R. 1970. "The alimentary canal of fish as an environment for helminth parasites." In: Taylor, A.E.R. and Muller, R. ed.: *Aspects of Fish Parasitology*. Symp. Brit. Soc. Paras. 8, Oxford and Edinburgh: Blackwell Sc. Publ., 43–77.
Willis, A.G. 1949. "On the vegetative forms and life history of *Chloromyxum thyrsites* Gilchrist and its doubtful systematic position." *Aust. J. Scient. Res.* (B). 2: 379–98.
Winn, H.E. 1955. "Formation of a mucous envelope at night by parrot fishes." *Zoologica* 40: 145–47.
Winn, H.E. and Bardach, J.E. 1959. "Differential food selection by moray eels and a possible role of the mucous envelope of parrot fishes in reduction of predation." *Ecology* 40: 296–98.
Winn, H.E. and Bardach, J.E. 1960. "Some aspects of the comparative biology of parrot fishes at Bermuda." *Zoologica* 45: 29–34.
Wolf, P.H. 1972. "Occurrence of a haplosporidan in Sydney Rock Oysters (*Crassostrea commercialis*) from Moreton Bay, Queensland, Australia." *J. Invert. Pathol.* 19: 416–17.
Wolf, P.H. 1976a. "Oyster diseases and parasites in Australian oysters." *Malacological Rev.* 6: 140.
Wolf, P.H. 1976b. "Occurrence of larval stages of *Tylocephalum* (Cestoda: Lecanicephaloidea) in two oyster species from Northern Australia." *J. Invert. Pathol.* 27: 129–31.
Wolf, P.H. 1977. "An unidentified protistan parasite in the ova of the Black-lipped Oyster, *Crassostrea echinata*, from Northern Australia." *J. Invert Pathol.* 29: 244–46.
Wolf, P.H. and Sprague, V. 1978. "An unidentified protistan parasite of the Pearl Oyster, *Pinctada maxima*, in tropical Australia." *J. Invert Pathol.* 31: 262–63.
Wootten, R. 1978. "The occurrence of larval *Anisakis* nematodes in small gadoids from Scottish waters." *J. mar biol. Assoc. U.K.* 58: 347–56.
Wootten, R. and Waddell, I.F. 1977. "Studies on the biology of larval nematodes from the musculature of cod and whiting in Scottish waters." *J. Conseil Inter. pour l'Exploration de la Mer.* 37: 266–73.
Wood, J.L. and Andrews, J.D. 1962. "*Haplosporidium costale* (Sporozoa) associated with a disease of Virginia oysters." *Science* 136: 710–11.
Wright, K.A. 1980. "Nematode sense organs." *Nematodes as Biological Models* 2: 237–95.

Wunder, W. 1929. "Die *Dactylogyrus* Krankheit der Karpfenbrut, ihre Ursache und ihre Bekämpfung." *Z. Fischerei* 27: 511–45.

Wynne-Edwards, V.C. 1962. *Animal Dispersion in Relation to Social Behaviour.* Edinburgh: Oliver and Boyd.

Yablokov, A.V., Bel'Kovich, V.M. and Borisov, V.I. 1972. (*Whales and Dolphins*) (Monographic outline). Moscow, USSR. Izd. "Nauka" (in Russian).

Yamaguti, S. 1958. Systema Helminthum. Vol. I. *Digenetic Trematodes.* Parts I and II. New York–London Interscience.

Yamaguti, S. 1959. Systema Helminthum. Vol. II. *Cestodes.* New York–London: Interscience.

Yamaguti, S. 1962. Systema Helminthum. Vol. III. *Nematodes.* Parts I and II. New York–London: Interscience.

Yamaguti, S. 1963a. Systema Helminthum. Vol. IV. *Monogenea and Aspidocotylea.* New York–London: Interscience.

Yamaguti, S. 1963b. Systema Helminthum. Vol. V. *Acanthocephala.* New York–London: Interscience.

Yamaguti, S. 1968. *Monogenetic Trematodes of Hawaiian Fishes.* Honolulu: University of Hawaii Press.

Yamaguti, S. 1970. *Digenetic Trematodes of Hawaiian Fishes.* Keigaku Publ. Co.

Yoshimura, K., Akao, N., Kondo, K. and Ohnishi, Y. 1979. "Clinicopathological studies on larval anisakiasis, with special reference to the report of extra-gastrointestinal anisakiasis." *Jap. J. Parasit.* 28: 347–54.

Young, M.R. 1938. "Helminth parasites of New Zealand. A bibliography with alphabetical lists of authors, hosts, and parasites." *Tech. Commun. Bur. Helminth. St. Albans.* 22: 1–19.

Young, M.R. 1939. "Helminth parasites of Australia. A bibliography with alphabetical lists of authors, hosts and parasites." *Tech. Commun. Bur. Helminth. St. Albans.* 23: 1–145.

Young, P.C. 1968. "Two new species of the family Allodiscocotylidae Tripathi, 1959 (Monogenoidea) from the gills of *Sphyraena obtusata* Cuvier and Valenciennes, with a note on the distribution of *Vallisiopsis australis* sp. nov." *J. Helminthol.* 42: 421–34.

Young, P.C. 1970. "The species of Monogenoidea recorded from Australian fishes and notes on their zoogeography." *An. Inst. Biol. Univ. Nal. Autón, México* 41, ser. zool. 1: 163–76.

Youngbluth, M.J. 1968. "Aspects of the ecology and ethology of the cleaning fish *Labroides phthirophagus* Randall." *Z. Tierpsych.* 25: 915–32.

Yoshino, T.P. 1976. "Encapsulation response of the marine prosobranch *Cerithidea californica* to natural infections of *Renicola buchanani* sporocysts (Trematoda: Renicolidae)." *Int. J. Paras.* 6: 423–31.

Zhukov, E.V. 1960. "(Endoparasitic worms of fish of the Sea of Japan and the South-Kuril shallow water.)" *Trud. Zool. Inst. Akad. Nauk SSSR* 28: 3–146 (in Russian).

Zschokke, F. 1933. "Die Parasiten als Zeugen für die geologische Vergangenheit ihrer Träger." *Forschungen und Fortschritte* 9: 466–67.

Index of Scientific and Common Names

Abra, 59
Abramis, 131
Abudefduf, 73, 84
Acanthella, 33, Fig. 34
Acanthobothrium, 95, 119
Acanthocephala, 10, 31, 33, 49, 50, 95, 102, 105, 106, 113, 140, 143, 148, 149, 164, Fig. 34
Acanthocotyle, 63
Acanthopagrus, 49, 117, Plate II
Acanthor, 33, Fig. 34
Acanthurus, 80
Acipenser, 21, 157, 169
Acoela, 21
Actinopyga, 64
Aggregata, 14, Figs. 6, 8
Albatross, 42
Albertia, 29, 31
Algae, 12, 35, 50, 73, 84
Allopseudaxinoides, 132
Amblyrhynchus, 73
Ammothea, Fig. 38
Amoeba, 12, 161
Amphibdella, 116
Amphibdelloides, 116
Amphilina, 26
Amphipoda, 26, 39, 47, 49, 84, Fig. 48
Amussium, 163
Amyloodinium, 12, 98
Ancyrocephalinae, 66, Plate IV
Anelasma, 37
Angiostrongylus, 173
Anguilla, 12, 103
Anisakis, 113, 125, 156, 165, 170, 172
Annelida, 10, 16, 31, 33, 35, 49, 161, Fig. 1
Anomalops, 97
Anoplodactylus, Fig. 38
Anoplura, 42

Ant, 72, 116
Antarctophthirus, 52
Anthobothrium, 101
Anthozoa, 10, 21
Anthuridea, 39
Antipatharians, 37
Apanteles, 50
Apicomplexa, 10, 14, 47, 48, 50, Figs. 1, 5, 6, 7, 8
Aplodinotus, 94, 114
Apogonidae, 74
Apophallus, 170
Aporocotyle, 109
Arachnida, 10, 35
Archosargus, 51
Arctic char, 156
Argulus, 37, 95, 115, Fig. 41
Arripis, Plate IV
Arrow worms, 12
Arthropoda, 10, 35, 67, 88, Fig. 1
Ascetospora, 10, 16, 47, 50, 51, 88, 98, 108, 161, 162, 163, 167, Figs. 1, 16
Ascidia, 16, 35, 39, 49, 63
Ascomyzontidae, 37
Ascothoracida, 37, 56, Fig. 42
Ascothorax, 37
Aspidoecia, 51
Aspidogaster, 108
Aspidogastrea, 23, 56, 61, 67, 90, 93, 108, 131, 147, 148, 149, 175, Fig. 24, Plate III
Aspidontus, 45, 85
Aspidophryxus, 51
Asterias, 16, 93, 98
Asteroidea, *see* starfish
Asteronyx, 45
Atherinops, 80, 169
Atlantic salmon, 115
Austrobilharzia, 59, 170, 171, Plate II

Austrogonoides, 42

Bacteria, 47, 50, 69, 73, 122
Barnacles, 37, 39, 56, 92, 93
Barracouta, 159
Bartholomea, 75
Bass, 76, 77, 129
Bdelloura, 23
Bearded seal (*see* seals)
Benedenia, 97, 98, 107, 118, 126
Bilharzia, 6
Biomphalaria, 88
Birds, 31, 33, 35, 42, 61, 62, 71, 72, 73, 86, 116, 151, 171, 172
Biting lice (*see* louse)
Bivagina, 98
Bivalvia, 11, 14, 23, 29, 33, 44, 48, 61, 84, 91, 92, 98, 105, 107, 108, 128, 160, 173
Blacklipped oyster (*see* oyster)
Blacksmith, 80
Blastodinidae, 8
Blastodinium, 48
Blenniidae, 45, 64, 85, 118
Blenny, 118
Bomolochus, 65
Bonellia, 44
Bopyridae, 39
Box, Fig. 13
Brachiopoda, 11, 35
Brachycoelum, 14
Branchiura, 37, 115, Fig. 41
Bream, 49, 117, Plate II
Brown algae (*see* algae)
Brown trout, 126
Bryozoa, 11, 84
Bucephalidae, 161
Bucephaloides, 59
Bucephalus, 51, 88, 91, 93, 161

Caddis fly, 39
Calanus, 48, Fig. 3
Calicotyle, 114
Caligoid copepods, 82, 83, 86
Caligus, 37, 51, 95, 112, 118, 123, 129, 133, Figs. 40, 62, Plates V, VI
Caliperia, 16, Fig. 10
Callichthyidae, 74
Callinectes, 164
Callorhynchus, Fig. 27
Campula, 165
Campulidae, 165

Cancricepon, 39, Fig. 45
Capelin, 65, 156
Caprellidae, 39, 84
Capsalidae, Plate IV
Carangidae, 74
Carangoides, Fig. 26
Caranx, 72
Carapus, 45, 65, Fig. 53
Carcinonemertes, 29, 64
Carcinus, 14, Fig. 43
Cardinal fish, 76
Carp, 122
Carpet shark (*see* shark)
Cat, 173
Catfish, 120, 122, 166
Centropristes, 76
Cephalopoda, 10, 14, 20, 26, 37, 39, 44
Cerastoderma, 16, 48, 91, 128, Fig. 9
Ceratioidea, 45
Ceratomyxa, 16, 47
Cercaria, 59, 61, 64, 66, 91, 92, 97, 118, 148, 170, 171, 172, Figs. 23, 58
Cerithidea, 89
Cerithium, 61, 90, 93, 108, 147, 148, Plate III
Cestoda, 10, 26, 29, 48, 49, 50, 54, 59, 63, 64, 65, 67, 68, 89, 90, 94, 95, 97, 101, 105, 106, 116, 118, 119, 120, 124, 131, 140, 143, 148, 149, 150, 156, 162, 164, 165, 170, Figs. 27, 28, 63, Plate I
Cestodaria, 26, 97, 123, 149, Fig. 27
Cetacea, 141, 143, 164, 165, 166, 171
Chaetodon, 81
Chaetodontidae, 74
Chaetognatha, 11, 12, 26, 49, 61, 65, 66, 93, Fig. 1
Chattonella, 48
Chimaera, 26, 149, Plates III, VI
Chloromyxum, 16, 159
Choerodon, 23, Plate I
Chordata, 11, 45, 49, Fig. 1
Chromis, 80, Fig. 59
Chrysophrus, 117
Chytridiopsis, 161
Cichlidae, 74, 83
Ciliates, 10, 16, 47, 48, 49, 50, 51, 63, 93, 98, 105, 125, 128, 161, 163, 164, 168, Figs. 1, 9, 10, 11, 12, 14, Plate VI
Ciliophora (*see* ciliates)
Ciona, 63
Cirripedia (*see* barnacles)
Clam, 29, 48, 51, 64, 90, 163, 169

Index of Scientific and Common Names   233

Clavella 112, 129
Cleaners, 72–86, 127, Fig. 59
Cleaner fish, 45, 73, 74, 75, 76, 77, 79, 80, 81, 82, 83, 84, 85, 86, Fig. 59
Cleaner shrimp, 73, 74, 75, 77, Plate VI
Clepticus, Fig. 59
Cliona, 20, 48, 113, 162
Clupea, 156, 158
Cnidaria, 10, 21, 35, 45, 48, 161, Figs. 1, 19
Coalfish, 63
Coccidia, 14, 158, 163, Figs. 5, 6, 7, 8
Coccidinium, 47
Cockle, 48
Cod, 49, 113, 125, 156, 159, 160, Fig. 71, Plates I, VI
Codworm, 160
Coelenterata, 44
Coeloplana, 21
Conidophrys, 51
Contracaecum, 31, 48, 113, 170, 172
Copepoda, 12, 26, 29, 35, 37, 39, 48, 49, 51, 61, 63, 65, 76, 77, 82, 86, 91, 92, 95, 98, 99, 100, 104, 105, 106, 107, 109, 110, 112, 113, 114, 115, 118, 119, 123, 124, 126, 129, 132, 133, 147, 148, 149, 150, 157, 158, 162, 163, 164, 166, 167, 174, Figs. 2, 3, 40, 70
Coracidium, 19, 59, 63
Corals, 7, 21, 39, 44, 45, 75, 77, 107, 116
Coral trout, 160, Plate VI
Corophium, 47
Corynosoma, 33
Coryphaenoides, Fig. 15
Crabs, 14, 21, 23, 26, 29, 31, 37, 39, 48, 50, 55, 64, 73, 74, 84, 92, 93, 161, 162, 163, 164, 170, Figs. 8, 21
Crangonobdella, Fig. 35
Crassicauda, 165, Fig. 55
Crassicutis, 51
Crassostrea, 48, 51, 88, 89, 91, 160, 162, Fig. 16
Crenilabrus, 72, 75, 76, 77, 79, 80, 81, 82, 83
Crenimugil, 118
Crepidula, 162
Crinoidea (*see* feather stars)
Crossopterygian fish, 148, 175
Crustacea, 11, 14, 23, 29, 31, 33, 35, 37, 39, 44, 48, 50, 56, 73, 74, 83, 84, 93, 95, 115, 162, 163, 164, 168, 172, 173,
Figs. 41, 42, 43, 44, 45, 46, 47, 48, Plate V
Cryptocaryon, 16
Crypocotyle, 92, 93
Cryptoniscina, 39, 50
Ctenolabrus, 83
Ctenophora, 10, 21, Figs. 1, 20
Cucullanus, 128, Fig. 61
Cunina, 21
Cunoctantha, 21
Cyamidae, 39
Cyamus, Fig. 48
Cybiidae, 26, 62
Cybiosarda, Plate II
Cyclophorus, 48
Cyclobothrium, 51
Cymbasoma, 35
Cymothoidae, 39
Cyprinidae, 74, 101, 151
Cyprinodontidae, 74
Cypris larva, 37
Cystacanth, 33, Fig. 34

Dactylogyrus, 114
Dactylozoites, Fig. 14
Dajidae, 39
Damsel fish, 73
Danalia, 39, 50, 56, Fig. 47
Decapoda, 39, 49, 74
Dendrogaster, 37, 56, Fig. 42
Deretrema, 97
Derogenes, 26, 125, 153, 154, Fig. 61
Diacolax, 44
Diadema, 76
Diaschistorchis, 56, Fig. 56
Dicentrarchus,129
Diclidophora, 63, 113, 126
Diclybothrium, 114
Dictyotyle, 63
Dicyema, Fig. 18, Plate I
Dicyemida, 10, 20, Fig. 18
Didymozoida, 52, Fig. 70, Plate II
Digenea, 23, 26, 54, 56, 67, 105, 106, 108, 136, 141, 144, 145, 147, 148, 149, 171, 178, Fig. 23
Dinoflagellida, 8, 12, 47, 48, Figs. 2, 3
Diphyllobothrium, 165, 170, 171
Diplectanum, 129
Diplogonoporus, 170, 171
Diplozoon, 112, 131
Discocotyle, 114

Dog, 171, 173
Dolphin, 39, 73, 164, 165, Plate II
Donax, 93
Dugong, 165

Earthworm, 31
Echeneibothrium, 64, 95, 118, 120
Echeneidae, 74, 82
Echinocephalus, 173
Echinodermata, 11, 16, 20, 23, 33, 35, 37, 44, 45, 49, Figs. 1, 42, 52
Echinoidea (*see* sea urchins)
Echinorhynchus, 113, 168
Echiurida, 11, 44, Fig. 1
Ectocotyla, 21, Fig. 21
Ectoprocta, 11, 35
Edriolychnus, Fig. 54
Edwardsia, 21
Eels, 12, 33, 103
Egretta, 171
Eimeria, 14, 158, Fig. 7
Elacatinus, 77, 81, Fig. 59
Elagatis, 72
Elasmobranchs, 35
Eleginus, 49
Ellobiopsis, 48, Fig. 3
Elytrophora, 95
Embiotocidae, 74
Endeis, Fig. 38
Endosphaera, 47
Enoplocotyle, 107
Entamoeba, 69
Entelurus, 82, 83
Enterohalacarus, 35
Enteropneusta, 11
Entobdella, 62, 63, 64, 119, 126, 131, Fig. 57
Entoconcha, Fig. 52
Entoconchidae, 44
Entoniscidae, 39
Entoprocta, 10, Fig. 1
Ephelota, 20, 164, Fig. 14
Epibdella, 120
Epinephelus, 81, Figs. 25, 59, Plates I, VI
Epitonium, 44
Ergasilus, 65, 95, 114, 126
Erignathus, 165
Eubothrium, 65, 156
Eucarida, 39
Eucestoda, 29, Fig. 28
Eudyptula, 42

Eulamia, Plate II
Eumetopias, 165
Eurysorchis, 124, Plate IV
Euthynnus, Plate IV

Fabespora, 51
Fabia, 92
Feather stars, 10, 33, 45,
Fecampia, 23, Fig. 22
Fecampiida, 23,
Fiddler crabs (*see* crab)
Finches, 73
Fish, 6, 7, 8, 12, 14, 16, 21, 23, 26, 29, 31, 33, 37, 39, 42, 45, 47, 49, 51, 52, 59, 61, 62, 63, 64, 65, 66, 71, 72, 73, 75, 76, 77, 79, 80, 81, 82, 83, 84, 85, 86, 87, 94, 95, 96, 97, 98, 100, 101, 102, 103, 104, 105, 106, 107, 108, 109, 110, 112, 113, 114, 115, 116, 117, 118, 119, 120, 122, 123, 125, 126, 128, 129, 130, 131, 132, 133, 136, 137, 139, 140, 141, 142, 143, 144, 145, 146, 147, 148, 149, 150, 151, 152, 153, 154, 155, 156, 157, 158, 159, 160, 166, 168, 169, 171, 172, 173, 174, 175, 176, Figs. 11, 53, 54, 64, 65, 66, 67, 69, 70,
Flagellates, 6, 8, 47, 48, 99, 161
Flatfish, 14
Flatworms, 21, 54
Flounder, 95, 110, 114, 119, 122, Fig. 61
Flukes, 6, 95, 96, 109, 120, 128, 171
Frigate birds, 42, 45
Frogs, 119
Fundulus, 76, 147
Fungi, 47, 50, 94
Fungia, 44
Fungicava, 44
Furnestia, 98

Gadidae, 150, 158
Gadus, 49, 113, 122, 125, 156, 158, 168, Fig. 64
Galapagos marine iguana, 73
Gametocytes, Figs. 6, 8
Gammarids, 84
Garibaldi, 80
Gasterosiphon, 44
Gasterosteus, 95, 120
Gastrocotyle, 65, 114, 115, 147
Gastrodes, 21
Gastropoda, 11, 16, 44, 48, 49, 84, Figs. 51, 52

Index of Scientific and Common Names 235

Gastrotricha, 10
Genocerca, 154
Genolinea, 154
Geospiza, 73
Gersemia, 45
Geryonia, 21
Gigantobilharzia, 172
Gillichthys, 114, 147
Glugea, 14, 113, 158
Gnathia, 39, 82, Fig. 44
Gnathiidea, 39, 83, 84, 115
Gnathostoma, 173
Gnathostomulida, 10
Gobiidae, 42, 74, 77, 81
Gononemertes, 29, Fig. 29
Gorgodera, 91
Gorgonians, 37, 81
Gorgonocephalus, 45
Gotocotyla, 26, 62
Grabham's cleaner shrimp, Plate VI
Grapsus, 73
Gregarines, 14, 47, 48, 161, 164
Grouper, 81
Gulls, 42, 61, 73, 171, 172, 177, Plate II
Gurnard, Plate VI
Gymnodinoides, 47
Gymnophallus, 14
Gyrocotyle, 26, 97, 123, Fig. 27, Plate VI
Gyrodactylidae, 62, 66, 95, 132
Gyrodactylus, 55, 94, 114, 119, 120, 131

Haddock, 116
Haemogregarina, 14, Fig. 5
Haemulidae, 74
Hagfish, 45
Hake, 102
Halarachne, 35
Half moon, 80
Halibut, 159
Halocercus, 165
Haplorchis, 170
Haplosporidium, 16, 162
Hemibdella, 35
Hemichordata, 11, 49, Fig. 1
Hemipera, 154
Hemiuridae, 125, 150
Hemiurus, Fig. 61
Heniochus, 84
Hermit crabs (*see* crabs)
Heron, 61, 171
Herring, 14, 156, 158, 159, 172
Heteromicrocotyla, Fig. 26

Heterophyes, 160, 170
Heterophyidae, 170, 171
Heteropneustes, 122
Hexabranchus, 75
Hexamita, 12, 99, 161, 170
Himasthla, 90, 92
Hippoglossus, 63
Hippolysmata, 77, Plate VI
Hirudinea (*see* leeches)
Histriobdella, 33, 73
Holothuria (*see* sea cucumber)
Holothuroidea, 11, 44
Homarus, 73
Hookworms, 31
Hoplitophrya, 48
Horsehair worms, 31
Horseshoe crab (*see* crab)
Hydractinia, 21
Hydrichthys, 21
Hydrobia, 114
Hydroids, 31
Hydrolagus, 123, Plates III, VI
Hydrozoa, 10, 21, 35, 44
Hyperia, 39
Hypocoma, 47
Hypocomella, 16, Fig. 9
Hypophthalmichthys, 114
Hyporhamphus, Fig. 64
Hypsypops, 80

Ichneumonidae, 50, 51
Ichthyosporidium, 48
Ichthyotomus, 33, Fig. 37
Inachus, 50
Insecta, 5, 6, 11, 39, 42, 50, 174, 177
Invertebrates, 5, 12, 14, 20, 23, 29, 31, 33, 37, 39, 57, 61, 62, 88, 90, 93, 94, 108, 116, 125, 169, 171, 172, 173, 175
Iophon, 20
Isopoda, 39, 49, 50, 51, 52, 56, 82, 83, 84, 106, 115, 149, 164, Figs. 44, 45, 46, 47, Plate V

John Dory, 83, 159

Kahawaia, Plate IV
"Katfisch", 166
Katsuwonus, 132, Plate III
Kinetoplastida, Fig. 4
Kingfish, 159, 160, Plate I
Kinorhyncha, 10
Kronborgia, 23, 55
Kudoa, 16, 158, 159, Plate I

Kuhnia, 129, 133, Fig. 62
Kyphosus, 102

Labridae, 72, 74, 77, 79, 82, 85
Labroides, 76, 77, 79, 80, 81, 82, 83, 84, 85, 86, Plate VI
Labyrinthormorpha, 10, 12, Fig. 1
Lagenorhynchus, 165
Lamellibranchia, 26, 49, 63
Lampreys, 45, 52
Lankesteria, 48
Larus, 171, Plate II
Latimeria, 148, 175
Laura, 37
Leather jacket, 83
Lecithaster, 108, Fig. 61
Leeches, 6, 10, 12, 14, 33, 35, 49, 100, 106, 113, 149, 164, Fig. 35
Lepas, 37
Lepeophtheirus, 51, 115, 119
Lepidapedon, 125
Leptococcus, 169
Leptonacea, 44
Leptonychotes, 52
Leptosynapta, 42
Lernaeenicus, 76
Lernaeid copepods, 83
Lernaeocera, 26, 37, 158, 166
Lernaeopodidae, 37, 115
Lernentoma, 129, 133
Leucothoe, 39
Limpet, 162
Limpet, false, 172
Limulus, 23
Linckia, 133
Liriopsis, 51
Littorina, 91, 93, 97, 114, 147
Liza, 118
Lobatostoma, 23, 61, 90, 93, 108, 131, 147, 175, Fig. 24, Plate III
Lobster, 14, 31, 33, 37, 73, 163, 164
Lophotaspis, 108
Lota, 150
Louse, 39, 42, 52
Lucernasia, Fig. 39
Lungfish, 83
Lungworm, 165

Mackerel, 26, 82, 110, 113, 132, 159, Plates II, V
Macoma, 98
Macrogametes, Fig. 8

Macrogyrodactylus, 83
Macrouridae, 148, 149, 150
Macrurus, 149
Malacobdella, 29, 48
Mallophaga, 42
Mallotus, 65, 156
Mammals, 6, 31, 42, 62, 72, 86, 96, 116, 164, 165
Marteilia, 100, 162
Mastigophora, 8, 10, 50
Mazocraes, 65
Medialuna, 80
Medusae, 21
Megadenus, 44
Meinertia, 51
Melanogrammus, 158, 166
Menestho, 161
Menidia, 65
Merlangius, 63, 65, 109, 112, 122, 156, 158
Merluccius, 102, 113
Merozoites, Fig. 8
Mesozoa, 10, 20, Figs. 1, 17, 18, Plate I
Metacercariae, 14, 48, 61, 68, 114, 148, 160, 164, Fig. 23
Metagonimus, 170
Metazoa, 10, 11, 96, 148,
Metchnikovellidae, 47
Mice, 70
Microcotyle, 94, 114
Microgametes, Fig. 8
Micromesistus, 113
Microspora, 10, 14, 16, 47, 48, 49, 50, 51, 95, 113, 149, 158, 160, 164, 166, Fig. 1
Minchinia, 16, 88, 98, 99, 100, 108, 161, 163, 167, 169, 170, Fig. 16
Miracidia, 59, 61, 63, Fig. 23
Mites, 35
Mnestra, 21
Mola, 52, 73
Mollusca, 6, 11, 14, 16, 20, 21, 23, 29, 35, 44, 48, 52, 55, 59, 63, 64, 67, 75, 84, 88, 89, 90, 91, 93, 96, 125, 160, 161, 175, Figs. 1, 50, 51, 52
Monkey, 173
Monodon, 165
Monogenea, 6, 10, 26, 49, 51, 54, 55, 57, 62, 63, 64, 65, 66, 67, 83, 87, 94, 95, 97, 98, 102, 105, 106, 110, 112, 113, 114, 115, 116, 117, 118, 119, 120, 123, 124, 125, 126, 129, 131, 132, 133, 136,

137, 139, 140, 141, 143, 144, 146, 147, 148, 149, 151, 152, 155, 157, 158, 174, Figs. 25, 60, 64, 65, 67, 68, 70, Plates IV, VI
Monopisthocotylea, 26, 115, 133, 146, Fig. 25, Plate IV
Monopterus, 122
Moray eel, 81
Morone, 95, 126
Mosquitoes, 6, 42
Mucronalia, 44
Mudskipper, 42
Mudworms, 33
Mugil, 118, 147, 157, 160, 169
Mullet, 97, 98, 118, 147, 157, 160, 169, Plate IV
Multicotyle, 56, 57
Mummichog, 147
Musculium, 91
Mussels, 16, 21, 39, 44, 61, 90, 91, 93, 98, 99, 100, 126, 147, 162, 163, 167, 170
Myoxocephalus, 168
Myriapoda, 11
Myriocladus, 56
Myrus, Fig. 37
Mysidacea, 39, 51,
Mytilicola, 61, 91, 92, 98, 99, 100, 107, 115, 126, 147, 162, 163, 167, 170,
Mytilus, 48, 91, 98, 99, 163
Myxidium, 16, Fig. 15
Myxine, 45
Myxobolus, 16, 157, 169
Myxozoa, 10, 14, 16, 47, 49, 51, 125, 149, 150, 157, 158, 159, 160, Figs. 1, 15
Myzopontiidae, 37
Myzostoma, Fig. 36
Myzostomida, 10, 23, 33, Fig. 36

Namakosiramiidae, 37
Nanophyetus, 170, 171
Narwhal, 165
Nasitrema, 165, Plate II
Nassarius, 92
Nauplius larvae, 35, 83
Nectonema, 31, Figs. 32, 33
Nemathelminthes, 10, 29, Fig. 1
Nemathobothrium, Plate II
Nematoda, 10, 31, 47, 48, 49, 50, 51, 52, 54, 57, 62, 70, 71, 95, 105, 106, 112, 113, 123, 125, 128, 137, 140, 141, 143,

148, 149, 150, 156, 159, 160, 163, 164, 165, 166, 168, 170, 172, 173, Figs. 30, 55
Nematomorpha, 10, 31, Figs. 32, 33
Nematopsis, 55, 161, 164
Nemertean, (*see* Nemertina)
Nemertina, 10, 16, 20, 29, 33, 48, 49, 64, 164, Figs. 1, 29
Neobrachiella, 129, 133
Neogrubea, 124
Neon goby, 81
Neopilina, 175
Neorhabdocoelida, 54
Neorickettsia, 171
Nicothöe, 164
Nitzschia, 157
Nosema, 14, 47, 51, 164
Notodelphyidae, 37
Nudibranchs, 73, 75

Obelia, Fig. 38
Oceanobdella, 113
Octocorallia, 44
Octodactylus, 113
Octopus, Figs. 6, 18, Plate I
Odostomia, 44, 107, 114
Oekiocolax, 48
Oenophorachona, Fig. 12
Oligochaeta, 10, 33
Oncomiracidium, 64
Onychophora, 11, Fig. 1
Oocyst, Figs. 7, 14
Oodinium, 12
Opalia, 44
Opalinata, 10, 12, 47, 50, Fig. 13
Opecoelina, Plate II
Ophichthidae, 45
Ophiolepis, 20
Ophiomaza, 45
Ophiuroidea, 11, 20, 33, 37, 45
Opisthobranchia, 44
Opisthotrema, 165
Orchitophrya, 16, 93, 98
Orectolobus, Plate I
Ornithobilharzia, 170
Orthohalarachne, 35
Orthonectida, 10, 20, Fig. 17
Osmerus, 158
Ostracoda, 39
Ostrea, 92, 160
Otodistomum, 113
Otostrongylus, 165

Oxycephalid amphipods, 82
Oxyjulis, 77, 79, 80, 84
Oysters, 6, 12, 16, 20, 31, 33, 44, 48, 51, 88, 89, 91, 92, 93, 98, 99, 100, 108, 113, 114, 115, 160, 161, 162, 164, 166, 167, 169, 170, 173, 177

Pachypygus, 63
Paedophoropus, 44
Pagurus, 26, 31
Paradinium, 48
Parafilaroides, 165
Paranthura, 39
Parasitosyllis, 33
Parorchis, 177
Parophrys, 113, 156
Parrot Fish, 72, 82
Parupeneus, 80
Pateriella, 39
Paucivitellosus, 64, 118
Peachia, 21
Pecten, 163
Pelecypoda (*see* bivalves)
Pelseneeria, 4
Peltogaster, 51
Penalla, 37
Penellidae, 37
Penguins, 14, 42
Pentastomida, 11, 42, Fig. 1
Percidae, 74
Periclimenes, 75
Peristernia, 90, 108, Plate III
Peroderma, 112, 158
Petrels, 42
Petromyzon, 45
Phanerodon, 79
Philanisus, 39
Philometra, 95, 113
Phoca, 151, 165
Phocaneme, 31, 113, 156, 160, 170, 172
Phoronidea, 11
Phoxichilidium, Fig. 39
Phtorophrya, 47
Phyllirrhoe, 21
Phyllobothrium, 95, Fig. 63
Pinctada, 162
Pinnipeds, 141, 143, 164, 171, 173
Pinnotheres, 64, 92, 162
Pinnotheridae, 39
Pipefish, 82, 83
Placophora, 11
Plaice, 75, 76, 82, 110, 119, 120, 122, 160

Planaxis, 61, 147, 171
Plasmodium, 14
Platichthys, 110, 122, 168
Platyhelminthes, 10, 21, 54, Fig. 1
Plectropomus, 160, Plate VI
Pleistophora, 14, 160
Plerocercoid, 26, 29
Plesiodiadema, 35
Pleuronectes, 110, 122
Podocotyle, 125, Fig. 61
Pogonophora, 11, 49, Fig. 1
Polar bear, 173
Pollachius, 63, 113
Polychaeta, 10, 14, 20, 33, 35, 47, 73, 74, 83, 113, 162, 166, Fig. 37
Polycladida, 48
Polydora, 33, 113, 162, 166
Polyopisthocotylea, 26, 115, 133, 146, Fig. 26, Plate IV
Polyp, 35, 44, 45
Polypodium, 21, Fig. 19
Polypterus, 83
Polystomoides, 57
Pomacentridae, 74
Pomphorhynchus, 95, Fig. 61
Pontobdella, 35
Porifera (*see* sponges)
Portunion, 39, Fig. 46
Portunus, 92, Fig. 8
Post-chalimus larvae, 115
Pouting, 63
Praniza larva, 76, 82, 83, Fig. 44, Plate V
Priapulida, 10, Fig. 1
Pricea, 26, 62
Proales, 31
Procercoid, 29
Proctoeces, 26, 163
Prosobranchia, 44, 133
Prosorhynchus, 125
Protancyrocephalus, 114
Protomyzostomum, 33
Protonymphon, 35
Protoopalina, 12, Fig. 13
Protozoa, 8, 10, 47, 48, 49, 50, 52, 55, 67, 69, 70, 88, 96, 98, 105, 106, 115, 147, 148, 149, 157, 162, 163, 164, 169, 170, Figs. 1-16
Psettarium, 109
Pseudaxine, 65, 114, 115, 147
Pseudochromidae, 74
Pseudocycnoides, Plate V
Pseudoklossia, 163

Pseudopleuronectes, 95
Pseudothoracocotyla, 113
Pterobranchia, 11
Pycnogonida, 11, 35, 84, Figs. 38, 39
Pygidiopsis, 170
Pyramidellidae, 44, 107, 114, 161
Pyrazus, 171

Radiolaria, 12, 47
Raja, 16, 63, 114, 118, 119, 120, Figs. 4, 5, 10
Rat, 70, 173
Ratfish, 123
Rays, 12, 63, 64, 73, 101, 114, 118, 119, 120
Red algae (*see* algae)
Rediae, 48, 68, 92, 93, Fig. 23
Red mullet, Plate IV
Reighardia, 42
Remora, 82
Renicola, 89
Reptiles, 42
Rhabdochona, 31
Rhabdocoela, 23
Rhabdosargus, 117
Rhizocephala, 37, 39, 51, 164, Fig. 43
Rhopalomenia, Fig. 50
Rhopalonema, 21
Rhopalura, 20, Fig. 17
Rockfishes, 109
Rotatoria, 10, 29, 31, Fig. 31
Rotifera, (*see* Rotatoria)
Roundworms, 31
Rugogaster, 175, Plate III
Rynkatorpa, 45

Sacculina, 37, 39, 51, 56, 92, 93, 164, Fig. 43
Sagitta, 26
Salarias, 118
Salmo, 114, 115, 126, 155, 156
Salmon, 115, 155, 156, 171, Plate IV
Salpa, 21, Fig. 20
Salvelinus, 156
Samson fish, Plate V
Sarcodina, 10, 12, 47, 48
Sarcomastigophora, 8, 10, Figs. 1, 2, 3, 4, 13
Sarcotaces, 115
Sardines, 112, 158
Scad, Plate V
Scallops, 163

Scaphopoda, 11, 44
Scaridae, 72, 74
Schistosomatidae, 61, 88, 170, 171, 172, Plate II
Sclerocrangon, Fig. 35
Scomber, 82, 129, 133, Figs. 62, 64, Plate II
Scomberomorus, 26, 113, 132, Plate V
Scorpaenidae, 109
Scyphidia, 168
Scyphomedusae, 39
Scyphozoa, 10
Sea anemones, 21, 29, 75
Sea cucumbers, 31, 42, 44, 45, 64, Fig. 53
Sea elephant, 165
Sea lion, 165
Seals, 31, 33, 35, 42, 52, 71, 151, 165, 173
Sea perch, 79
Sea snakes, 42
Sea spiders, 35
Sea stars (*see* starfish)
Sea urchins, 11, 35, 44, 73, 76
Sebastes, 109, 128
Seison, 31
Señorita, 77, 79, 80, 84
Sepia, 26, Fig. 8
Seriola, 159, Plates I, V
Seriolella, 123, Plate IV
Serranus, 81
Shad, 65
Sharks, 12, 26, 29, 37, 72, 105, Plates I, II
Shellfish, 100
Shovel-nose ray, 83
Shrimps, 23, 73, 74, 75, 77, 163, 164, 171
Sillago, 132
Silver gull (*see* gull)
Siphamia, 76
Siphonaria, 172
Siphonostomatoida, 37
Sipunculoids, 11, 23, 44, Fig. 1
Skates, 23, 64
Skipjack tuna, 132, Plate III
Skuas, 45
Smelt, 158, 160
Snails, 5, 21, 23, 33, 44, 61, 88, 89, 90, 91, 92, 93, 97, 107, 108, 114, 127, 133, 147, 161, 162, 169, 171, 175, 177, Fig. 52

Snake eels, 45
Snapper, 117
Snoek, 159
Sole, 35, 113, 126, 131, 156
Solea, 62, 63, 119, 126, 131
Solenogastres, 11, 44, Fig. 50
Sparidae, 117
Spelotrema, 14
Sphenophrya, 161
Spiny headed worms (*see* Acanthocephala)
Spirochaetes, 50
Spirometra, 170
Spisula, 51, 163
Sponges, 10, 20, 35, 39, 48, 113, 162, Fig. 1
Spores, 14, 62, 163, Figs. 7, 8, 16
Sporocysts, 48, 68, 91, 92, 93, Fig. 23
Sporozoites, 14, Figs. 7, 8
Sprat, 158
Sprostonia, Fig. 25
Squid, 48, 172
Staphylorchis, Plate II
Starfish, 11, 16, 37, 39, 42, 44, 56, 93, 98, 133, Fig. 42
Stellantchasmus, 170
Stenurus, 166
Stockfish, 159
Sturgeon, 21, 26, 114, 157, Fig. 19
Stylifer, 44
Stylochus, 100
Sucking lice (*see* louse)
Suctoria, 47, 164, Fig. 14
Sulcascaris, 163
Sunfish, 52, 73
Surf clams (*see* clams)
Surgeon fish, 80
Symphodus, 77, 80
Syncoelium, 23, 124, Plate III
Syndinium, 48, Fig. 2
Syngnathidae, 74
Syngnathus, 82

Tachyblaston, 16, Fig. 14
Tachysurus, 120
Taeniura, 101
Tapes, 90
Tapeworms (*see* Cestoda)
Tardigrada, 11, 42, Figs. 1, 49
Tarwhine, 117
Teleostei, 12, 29, 97, 122, 124, 140, Figs. 65, 67

Tentaculata, 11, 74, Fig. 1
Terns, 42
Terranova, 172
Tetrakentron, 42, Fig. 49
Tetraphyllidae, 148, Fig. 61
Tetraphyllidean larvae, Fig. 61
Tetrarhynchidae, 105
Thais, 177
Thalassoma, Fig. 59
Thelohania, 164
Thigmotrichid ciliates (*see* ciliates)
Thoracica, 37
Thorny-headed worms (*see* Acanthocephala)
Thyca, 44, 133, Fig. 51
Thynnascaris, 31, 113, 159, 168, 172, Figs. 30, 61, 71
Thynnus, 95
Thyrsites, 159
Thysanozoon, 48
Ticks, 35, 73
Tilapia, 83
Tintinnoinea, 47
Topminnow, 76, 77
Top smelt, 80
Trachinotus, 23, 61, 131, 175
Trachurus, 65, 114, 115, Fig. 64, Plate V
Transversotrema, 23, 64
Trematoda, 10, 12, 47, 48, 49, 50, 51, 52, 54, 56, 59, 61, 63, 65, 66, 67, 68, 88, 89, 90, 91, 92, 93, 96, 97, 102, 103, 108, 113, 114, 116, 118, 119, 124, 125, 127, 128, 131, 136, 137, 139, 140, 141, 142, 143, 147, 149, 150, 151, 152, 153, 154, 157, 159, 161, 163, 164, 165, 169, 170, 171, 175, 177, Figs. 23, 60, 68
Triaenophorus, 94
Trichinella, 70, 170, 173
Trichodina, 16, 47, 48, 98, 168, Fig. 1
Tricladida, 23, 48
Trigla, 112, 116, 123, 129, 133, Plate VI
Trisopterus, 63
Trochopus, 115, 133
Trypanorhyncha, 29, Fig. 28, Plate I
Trypanosoma, 12, 70, Fig. 4
Trypanosomiasis, 6
Tubulovesicula, 154
Tuna, 95, Plates III, IV
Tunicates, 29, 37
Turbellaria, 10, 20, 21, 47, 48, 49, 54, 161, 164, Figs. 21, 22
Turritella, 92
Turritopsis, 21

Turtles, 23, 56, 57, 73, 168
Tuskfish, 23, Plate I
Tylocephalum, 90, 162

Uca, 31
Udonella, 51, 118, Plate VI
Unicapsula, 159, 160
Upeneichthys, Plate IV
Uranotaenia, 42
Urosporidium, 51, 163, 164

Vahlkampfia, 161
Velacumantus, 169
Venerupis, 64
Vertebrates, 5, 23, 31, 33, 42, 45, 49, 61, 86, 88, 89, 94, 96, 108, 173, 175

Walrus, 173
Warehou, Plate IV
Weddel seal (*see* seal)
Whale lice (*see* louse)

Whales, 37, 39, 52, 71, 164, 165, 166, 173, Fig. 55
White-sided dolphin (*see* dolphin)
Whiting, 63, 65, 109, 112, 122, 132, 156
Wrasses, 72, 76, 77, 79, Plate VI

Xiphosura, 11

Zanclea, 21
Zebrasoma, 81
Zelinkiella, 31, Fig. 31
Zeus, 83
Zirfaea, 29
Zoarces, 168
Zoochlorella, 47
Zoogonoides, Fig. 61
Zoogonus, 113
Zooxanthella, 47
Zostera, 12
Zygocotyle, 131

Subject Index

Abscess. *See* tissue reactions
Age of host, effect on parasites, 66, 94, 105, 114-15, 123, 126-27, 149, 155, 158, 163
Age of oceans, effect on parasites, 154, 155
Aggregation of parasites, 65, 66. *See also* distribution
Anisakiasis, 62, 172, 173
Anting, 72
Asexual reproduction, 54, 66, 67, 68, 133

Beaching, 165, 166
Behaviour of hosts in relation to parasite infection, 63, 64, 65, 66, 91, 92, 93, 165, 166. *See also* cleaning symbiosis
Benefit of parasite to host, 69-71
Bilharzia, 6
Biological markers, parasites as b.m., 155, 156

Carrying capacity, 54, 97
Character displacement, 129
Chronic inflammation. *See* tissue reactions
Clam diggers itch. *See* dermatitis
Cleaner mimic. *See* cleaning symbiosis
Cleaning symbiosis, 72-86, 127, Fig. 59, Plate VI; cleaner mimic, 45, 85, 86; cleaning station, 77, 79, 84; dancing of cleaner fish, 76, 85; ecological function of c.s., 82, 83, 85-86, 176; food of cleaners, 73, 75, 77, 82-84, 86; guild signs of cleaners, 77, 85; host specificity of cleaners, 79; invitation postures of host fish, 79, 80, 81, 82, Fig. 59; latitudinal gradients in c.s., 84-85; pseudocleaning, 81; signal markings of cleaners (*see* guild signs); site specificity of cleaners, 79
Coevolution of hosts and parasites, 69, 100
Commensalism, 5, 21, 29, 31, 33, 39, 44, 45, 69, 73, 161, 166
Competition, intraspecific, 96, 130, 131, interspecific, 104, 116, 128, 129. *See also* niche
Competitive exclusion, 116, 128. *See also* niche
Complexity of parasites, 49, 54, 55-59, 101, 131
Continental drift, 154
Control of marine parasites,168-70, 173, 176, 177
Cross-fertilization of parasites, 67, 131
Cultures, effect of parasites on host c., 160, 167, 168

Density, of host populations, 67, 96, 98, 99, 100, 107, 131, 141, 163, 168
Density, of infection, 107, 108
Density, of parasite population, 68, 98, 127, 130, 132, 133
Density dependence, 96, 97
Density independence, 96
Depressions. *See* zoogeography
Depth of water, 52, 99, 113, 125, 148-50, 153, 154. *See also* zoogeography
Dermatitis, 171, 172
Diet of hosts, effect on parasites, 66, 125, 127, 149, 150, 168
Dispersal of parasites, 59-61, 66, 67, 98, 99, 102, 103, 177, Fig. 58
Distribution: aggregated = clustered = contagious = overdispersed, 65, 66, 156; even = underdispersed, 65, 156; random, 65, 66

Subject Index 243

Divergence rule. *See* Eichler's rule
Drugs for marine parasites, 168
Eichler's rule, 101
Endemicity. *See* zoogeography
Endogenous rhythms, 62, 63, 119
Enhancement of mating chances. *See* niche
Epidemics. *See* epizootics
Epizootics, 96, 98, 100, 157, 167, 168, 169

Facultative parasites, 4, 99
Fahrenholtz's rule, 101
Fecundity of parasites. *See* number of offspring
Fluctuations in parasite infections, 69, 115, 147, 156
Food of parasites. *See* niche
Frequency of infection, 94, 96, 97, 107, 108, 114, 125, 128, 131, 132, 141, 143, 144, 149, 150, Fig. 67
Fundamental niche. *See* niche

Generation time, 55, 69, 152
Geographical isolation. *See* speciation
Geographical range. *See* niche
Gigantism, 93
Gradients. *See* zoogeography
Granuloma. *See* tissue reactions
Guild sign. *See* cleaning symbiosis

Haemorrhage. *See* tissue reactions
Hatching of parasites, 26, 62, 63, 118, 119, 126
Hermaphroditism, 67, 68, 131; sequential, 67; simultaneous, 68
Herring worm disease, 172
Hexamitiasis, 99, 161
Host range, 107, 108, 144, Fig. 68 (*see also* host specificity and niche)
Host specificity, 101, 105-8, 118, 119, 120, 150; ecological h.s., 105, 118; h.s. of cleaners (*see* cleaning symbiosis); per cent specificity, 108; phylogenetic h.s., 101, 105; specificity index, 107, 108, 144, Fig. 60; (*see also* niche and host range)
Hyperparasites, 4, 12, 14, 47, 50, 51, 56, 105, 117, 118, 127, 171, Plate VI
Hyperplasia. *See* tissue reactions

Immunity, 72, 86, 87, 96; adaptive i., 86, agglutination, 88; antibody 87, 88, 120; antigen, 87, 120; essential antigen, 87; functional antigen (*see* essential antigen); immunization, 66; immunoglobulin, 88, 120; non-self, 86, 87, 88; protective antibody, 87; self, 86, 87; serum protein, 88. *See also* resistance
Importance of parasites, 1, 5, 6, 7, 16, 37, 157, 158, 159, 160, 161, 162, 163, 164, 166-67, 177
Incidence. *See* frequency of infection
Index parasites. *See* zoogeography
Infection mechanisms, 12, 14, 20, 21, 23, 26, 29, 31, 33, 59, 61-65, 66, 67, 87, 98, 118, 119, 131, Figs. 8, 14, 30
Instability of host-parasite systems, 96
Intensity of infection, 65, 94, 95, 96, 97, 99, 107, 125, 128, 131, 132, 141, 143, 144, 161, 163, 168
Interactive site segregation. *See* niche
Intraspecific parasites, 4, 44, 45, Fig. 54
Intrinsic rate of population growth, 54
Invitation posture. *See* cleaning symbiosis

Kleptoparasites, 45
K-selection, 54

Larval parasites, 4, 39
Latent parasites, 4
Latitudinal gradients. *See* cleaning symbiosis and zoogeography
Leitparasiten. *See* index parasites
Leucocytosis. *See* tissue reactions
Limiting resource. *See* niche
Littoral parasites. *See* zoogeography
Living fossils, 175

Macrohabitats. *See* niche
Malaria, 6
Mass mortalities, 6. 16. 33. 95. 96-100, 157, 158, 160, 161, 162, 163-64, 167, 176
Mating hypothesis, 66, 131-35. *See also* niche
Meningoencephalitis, eosinophilic, 173
Metaplasia. *See* tissue reactions
Microhabitats. *See* niche
Microhabitat width. *See* niche
Migrations of hosts, 12, 65, 118, 125, 155, 156
Mucus envelope, 72
Mud blisters, 166
Multidimensional hyperspace. *See* niche

Mutation rates, 68, 152
Mutualism, 5, 21, 69, 70
Nacreazation. *See* pearl formation
Necrosis. *See* tissue reactions
Nervous system of parasites, 55, 56, 57, 59, Fig. 56
Niche, 104, 105, 118; adaptions to n., 52; biological function of n. restriction, 118, 127-35; causes of n. restriction, 118-27, 176; character displacement and n. segregation (*see* character displacement); competition effect on n., 104, 116, 127, 128, 129, 130, 147, 176 (*see also* removal experiments); n. dimensions, 104. (*see also* sex of host, age of host, season, hyperparasitism); empty n., 117, 130, 146, 176; enhancement of mating chances by n. restriction, 66, 127, 131, 132, 133, 135, 176, Fig. 64 (*see also* mating hypothesis); n. expansion, 127,130; extrinsic factors affecting n., 127; finding of n., 59, 123; food as n. dimension, 104, 115-16, 130, 144, 146; fundamental n., 104, 116, 128, 144, 147; geographical range as n. dimension, 105, 113-14, 124-26; host range and host specificity as n. dimensions, 105, 118, 144, Figs. 63, 68, (*see also* host specificity); interactive site segregation, 128; intraspecific (intrinsic) factors affecting n., 127, 130-35; limiting resource, 129; macrohabitats as n. dimension, 105, 112-13, 118, 124-26; mating hypothesis, 132, 133, 135; microhabitats as n. dimension, 97, 104, 105, 108-12, 117, 118, 120-24, 127, 129, 130, 131, 132, 133, 135, 144, 145, 174, 176, Figs. 61-64, 69, 70; multidimensional hyperspace, 104; number of niches, 117; proximate causes of n. restriction, 118; realized niche, 104, 116, 128; reinforcement of reproductive barriers and n. segregation, 127, 128, 129, 130, 147, 176; saturation of n., 97, 116-18, 176; selective site segregation, 128; n. width, 144, 176; ultimate causes of n. restriction, 118, 127-35
Non-self. *See* immunity

Number of offspring, 23, 54, 55, 68, 96, 97, 131, 157, 158
Number of parasitic species, 5, 6, 7, 8, 10, 11, 23, 37, 42, 48, 49, 50, 54, 55, 96, 98, 117, 125, 137, 141, 142, 143, 144, 148, 149, 150, 152, 154, 155, 157, 174

Obligatory parasites, 4
Overfishing for parasite control, 169
Overinfection, 59

Parasitic castration, 14, 16, 93, 94, 97, 158, 161, 163
Parasitoids, 6, 50
Paratenic hosts, 5, 33, 62
Parthenogenesis, 54, 67, 68
Partitioning of microhabitats. *See* niche
Pathogenicity of parasites, 96, 100, 161, 162, 167
Pathological effects, 90-95, 157-67, 170, 171, Fig. 71
Pearl formation, 89
Pepper crab disease, 164
Performance of infected hosts, 92
Periodic parasites, 4
Permanent parasites, 4
Phagocytosis. *See* tissue reactions
Phoresis = phoresy, 5, 69
Pit disease. *See* hexamitiasis
Population size of parasite, 54, 68, 157. *See also* density
Progenesis, 26, 68
Protrandy, 68
Proximate causes of niche restriction. *See* niche
Pseudocleaning. *See* cleaning symbiosis
Pus. *See* tissue reactions

Regulatory mechanisms, 96, 97, 127
Reinforcement of reproductive barriers. *See* niche
Relative species diversity of parasites, 140, 141, 152, 155, Fig. 65
Relict fauna. *See* zoogeography
Removal experiments, 85, 86, 104
Reproductive barriers. *See* niche
Reproductive capacity. *See* number of offspring
Reproductive potential. *See* number of offspring
Resistance to parasites, 66, 88, 90, 92, 99, 160, 167, 169. *See also* immunity

Resorption of dead parasites. *See* tissue reactions
Retardation of growth, 94, 158, 159, 160, 161, 162
Retarded evolution of parasites, 101
r-selection, 54
Salinity, 99, 113, 125, 169
Salmon poisoning disease, 171
Saturation of niches. *See* niche
Sea bathers' eruption. *See* dermatitis
Season, effect on parasites, 105, 115, 126, 147, 148, 155, 170
Seasonal fluctuations, 66
Selective site segregation. *See* niche
Self. *See* immunity
Self-fertilization of parasites, 67, 131
Sense organs of parasites, 39, 45, 55, 56, 57, 59, Fig. 57
Sex of host, effect on parasites, 105, 114, 123, 126
Signal markings. *See* cleaning symbiosis
Site specificity. *See* niche
Size of parasites, 4, 8, 49, 52, 68, Fig. 55
Speciation, 174, 175; allopatric = geographical = s. by geographical isolation, 154, 174; sympatric, 174
Specificity index. *See* host specificity
Speed of evolution, 68, 128, 152, 175, 176
Starvation autolysis, 91
Stranding. *See* beaching
Supercooling, 52
Susceptibility, 66, 97, 100, 120, 168
Swimmer's itch. *See* dermatitis
Symbiosis, 5, 47, 69
Szidat's rule, 101

Temperature, effect on parasites, 1, 52, 104, 124, 125, 126, 142, 147, 150, 151, 152, 153, 154, 156, 161, 175, 176, Figs. 65, 66, 67, 68
Temperature stress, 90, 92

Temporary parasites, 4, 6, 37
Tissue reactions, 72, 86, 88, 89, 90, 95, 165; abscess, 89, 167; autolysis, 91; brown cells, 89, 90; capsule formation (*see* encapsulation); chronic inflammation, 87; encapsulation, 87, 89, 90; fibrosis, 90, 91, 95, Plate III; granuloma, 95; haemorrhage, 95; hyperplasia, 95; inflammation, 86, 87, 89; leucocytosis, 89; metaplasia, 91; necrosis, 87, 89, 91; oedema, 87, 89; phagocytosis, 86, 87, 88, 89, 90; pus, 89; resorption of dead parasites, 90; vasodilation, 87, 89. *See also* pearl formation
Transport host. *See* paratenic host
Trypanosomiasis, 6

Ultimate causes of niche restriction. *See* niche

Vasodilation. *See* tissue reactions
Verminous pneumonia, 165
Von Ihering method. *See* zoogeography

Weed, itch. *See* dermatitis
White spot disease, 16

Zoogeography of marine parasites, 2, 136-58, 175-76; bathypelagic zone, 45, 149; benthopelagic zone, 149, 150; cosmopolitan species, 153; deep water, 148, 149, 150, 151, 152, 153, 175; depressions, 153; differences between Atlantic and Indo-Pacific, 137, 139, 140, 151, 154, Figs. 65, 66; endemicity, 137, 151, 152, 154, 155; index parasites, 102; latitudinal gradients, 6, 125, 130, 140-47, 150, 151, 152, 153, 175, Figs. 65-68; littoral parasites, 125; relict fauna, 150, 151; remote oceanic islands, 151, 152; von Ihering method, 102; zoogeographical regions, 102, 136, 137, 141.